Simulation Lab Manual

for use with

Electricity for the Trades

Third Edition

Prepared by
Frank D. Petruzella

Mc Graw Hill Education

SIMULATION LAB MANUAL FOR USE WITH

ELECTRICITY FOR THE TRADES, THIRD EDITION

Published by McGraw-Hill Education, 2 Penn Plaza, New York, NY 10121.

Some ancillaries, including electronic and print components, may not be available to customers outside the United States.

This book is printed on acid-free paper.

1 2 3 4 5 6 QVS 23 22 21 20 19

ISBN 978-1-260-43737-9
MHID 1-260-43737-X

Cover image: ©Chockchai Paralart/123RF.

mheducation.com/highered

Electricity For The Trades, 3/e

Paperless Multisim Simulation Lab Manual

Table of Contents

SECTION 1 Fundamentals of Electricity

	Topic	Lab Assignment	File at Mhhe.com/EFT3E	Pg no.
Ch. 1	**Safety**	- Electric shock	**CH 01_01**	**3**
Ch. 2	**Atoms and Electricity**	- Continuity test circuit	**CH 02_01**	**4**
Ch. 3	**Sources and Characteristics of Electricity**	- DC and AC current electricity	**CH 03_01**	**5**
Ch. 4	**Electrical Quantities and Ohm's Law**	- Ammeter connection	**CH 04_01**	**6**
		- Voltmeter connection	**CH 04_02**	**7**

		- Ohmmeter connection	CH 04_03	8
		- Effect of voltage on current flow	CH 04_04	9
		- Wattmeter connection	CH 04_05	10
		- Ohm's Law (1)	CH 04_06	11
		- Ohm's Law (2)	CH 04_07	12
Ch. 5	**Simple, Series, and Parallel Circuits**	- Simple circuit	**CH 05_01**	**13**
		- Series circuit	**CH 05_02**	**14**
		- Parallel circuit	**CH 05_03**	**15**
		- Pilot lamp circuit	**CH 05_04**	**16**
		- Series/parallel devices	**CH 05_05**	**17**
		- Series circuit troubleshooting	**CH 05_06**	**18**

- Series circuit analysis (1)	CH 05_07	19
- Series circuit analysis (2)	CH 05_08	20
- Parallel circuit analysis (1)	CH 05_09	21
- Parallel circuit analysis (2)	CH 05_10	22
- Series circuit analysis (3)	CH 05_11	23
- Parallel circuit analysis (3)	CH 05_12	24
- Control devices analysis (1)	CH 05_13	25
- Control devices analysis (2)	CH 05_14	26
- Control devices analysis (3)	CH 05_15	27

		- Control devices analysis (4)	CH 05_16	28
		- Open and shorted faults	CH 05_17	29
Ch. 6	**Measuring Voltage, Current, and Resistance**	- Determining polarity	**CH 06_01**	**30**
		- Circuit continuity	**CH 06_02**	**31**
		- Test lead connections	**CH 06_03**	**32**
		- Ground-referenced voltages	**CH 06_04**	**33**
		- Multimeter troubleshooting (1)	**CH 06_05**	**34**
		-Multimeter troubleshooting (2)	**CH 06_06**	**35**
		- Power supply measurements (1)	**CH 06_07**	**36**

- Power supply measurements (2)	**CH 06_08**	**37**
- Series circuit measurements (1)	**Ch 06_09**	**38**
- Current measurement	**CH 06_10**	**39**
- Resistance measurement	**CH 06_11**	**40**
- Component troubleshooting	**CH 06_12**	**41**
- Branch circuit current measurement	**CH 06_13**	**42**
- Series circuit measurements (2)	**CH 06_14**	**43**
- Parallel circuit measurements	**CH 06_15**	**44**

Ch. 7	**Ohm's Law**	- Calculating current (1)	**CH 07_01**	**45**
		- Current/resistance relationship	**CH07_02**	**46**
		- Calculating voltage (1)	**CH 07_03**	**47**
		- Calculating resistance (1)	**CH 07_04**	**48**
		- Current/voltage relationship	**CH 07_05**	**49**
		- Calculating power	**CH 07_06**	**50**
		- Calculating voltage (2)	**CH 07_07**	**51**
		- Calculating current (2)	**CH 07_08**	**52**
		- Calculating resistance (2)	**CH 07_09**	**53**

		- Volts/Amps/Watts relationship	CH 07_10	54
		- Ohm's Law troubleshooting	CH 07_11	55
Ch. 8	**Resistors**	- Variable resistor	CH 08_01	56
		- Rheostat	CH 08_02	57
		- Potentiometer	CH 08_03	58
		- Series resistors	CH 08_04	59
		- Voltage-divider	CH 08_05	60
		- Series/ parallel resistors (1)	CH 08_06	61
		- Current-divider	CH 08_07	62
		- Series/ parallel resistors (2)	CH 08_08	63
		- Series/ parallel resistors (3)	CH 08_09	64

		- Series/ parallel resistors troubleshooting	CH 08_10	65
Ch. 9	Electricity and Magnetism	- Step-down transformer	CH 09_01	66
		- Control relay	CH 09_02	67
Ch. 10	Electrical Power and Energy	-Calculating power (1)	CH 10_01	68
		- Power/current calculation (1)	CH 10_02	69
		- Power/resistance calculation	CH 10_03	70
		- Power/current calculation (2)	CH 10_04	71
		- Calculating power (2)	CH 10_05	72
		- Calculating energy	CH 10_06	73

SECTION 2 Direct Current (DC) Circuits

	Topic	Lab Assignment	File Mhhe.com/EFT3E	Pg no.
Ch. 11	Solving the DC Series Circuit	- Circuit characteristics	CH 11_01	77
		- Calculating E, I, and R (1)	CH 11_02	78
		- Calculating E, I, and R (2)	CH 11_03	79
		- Circuit analysis (1)	CH 11_04	80
		- Circuit analysis (2)	CH 11_05	81
		- Voltage measurement	CH 11_06	82
		- Aiding and opposing voltages	CH 11_07	83
		- Voltage source resistance	CH 11_08	84

		- Series conductor resistance	**CH 11_09**	**85**
		- Circuit troubleshooting (1)	**CH 11_10**	**86**
		- Circuit troubleshooting (2)	**CH 11_11**	**87**
		- Circuit troubleshooting (3)	**CH 11_12**	**88**
		- Circuit troubleshooting (4)	**CH 11_13**	**89**
		- Oven control circuit	**CH 11_14**	**90**
Ch. 12	**Solving the DC Parallel Circuit**	- Circuit characteristics	**CH 12_01**	**91**
		- Calculating E, I, and R (1)	**CH 12_02**	**92**
		- Calculating E, I, and R (2)	**CH 12_03**	**93**
		- Circuit analysis (1)	**CH 12_04**	**94**
		- Circuit analysis (2)	**CH 12_05**	**95**

		- Voltage polarity	**CH 12_06**	**96**
		- Circuit troubleshooting (1)	**CH 12_07**	**97**
		- Circuit troubleshooting (2)	**CH 12_08**	**98**
		- Circuit troubleshooting (3)	**CH 12_09**	**99**
		- Circuit solving (1)	**CH 12_10**	**100**
		- Circuit solving (2)	**CH 12_11**	**101**
		- Circuit solving (3)	**CH 12_12**	**102**
		- Circuit solving (4)	**CH 12_13**	**103**
Ch. 13	**Solving the DC Parellel Series-Parallel Circuit**	- Circuit characteristics	**CH 13_01**	**104**
		- Kirchhoff's Laws	**CH 13_02**	**105**
		- Circuit analysis	**CH 13_03**	**106**
		- Circuit troubleshooting (1)	**CH 13_04**	**107**

		- Circuit troubleshooting (2)	CH 13_05	109
		- Circuit troubleshooting (3)	CH 13_06	110
Ch. 14	Network Theorems	- Practical voltage source	CH 14_01	111
		- Practical current source	CH 14_02	112
		- Superposition Theorem (1)	CH 14_03	113
		- Superposition Theorem (2)	CH 14_04	114
		- Superposition Theorem (3)	CH 14_05	115
		- Superposition Theorem (4)	CH 14_06	116
		- Superposition Theorem (5)	CH 14_07	117

		- Thevenin's Theorem (1)	**CH 14_08**	**118**
		- Thevenin's Theorem (2)	**CH 14_09**	**119**
		- Thevenin's Theorem (3)	**CH 14_10**	**120**
		- Norton Theorem (1)	**CH 14_11**	**121**
		- Norton Theorem (2)	**CH 14_12**	**122**

SECTION 3 Alternating Current (AC) Circuits

	Topic	Lab Assignment	File Mhhe.com/EFT3E	Pg no.
Ch. 15	**Alternating Current Fundamentals**	- Time period	**CH 15_01**	**125**
		- Peak voltage	**CH 15_02**	**126**
		- RMS voltage	**CH 15_03**	**127**

		- Bridge rectifier	CH 15_04	128
		- Distribution transformer	CH 15_05	129
		- Three-phase system (1)	CH 15_06	130
		- Three-phase system (2)	CH 15_07	131
		- AC series resistive circuit	CH 15_08	132
		- AC parallel resistive circuit	CH 15_09	133
		- Three-phase resistive circuit (1)	CH 15_10	134
		- Three-phase resistive circuit (2)	CH 15_11	135
Ch. 16	**Inductance and Capacitance**	- RL time constant	CH 16_01	136
		- Inductance DC/ AC circuits	CH 16_02	137

- Inductance / current flow CH 16_03 138

- Inductance / frequency CH 16_04 139

- Inductive reactance CH 16_05 140

- Inductors series connected CH 16_06 141

- Inductors parallel connected CH 16_07 142

- Inductor / phase relationship CH 16_08 143

- Inductor / wattage CH 16_09 144

- Calculating inductive reactance CH 16_10 145

- Inductors series connected X_L CH 16_11 146

- Inductors parallel connected X_L CH 16_12 147

- Inductive Volt/ Amps Reactive	**CH 16_13**	**148**
- Capacitor charge/ discharge	**CH 16_14**	**149**
- RC time constant	**CH 16_15**	**150**
- Capacitance DC/ AC circuits	**CH 16_16**	**151**
- Capacitance / current flow	**CH 16_17**	**152**
- Capacitance / frequency	**CH 16_18**	**153**
- Capacitive reactance	**CH 16_19**	**154**
- Capacitors series connected	**CH 16_20**	**155**
- Capacitors parallel connected	**CH 16_21**	**156**
- Capacitor / phase relationship	**CH 16_22**	**157**

		- LC parallel circuit	CH 16_23	158
		- Capacitor / wattage	CH 16_24	159
		- Capacitive reactance	CH 16_25	160
		- Capacitive Volt/ Amps Reactive	CH 16_26	161
Ch. 17	Resistive, Inductive, Capacitive (RLC) Series Circuits	- R circuit	CH 17_01	162
		- RL circuit	CH 17_02	163
		- Adding and opposing voltages	CH 17_03	164
		- Triangular vector addition	CH 17_04	165
		- Parallelogram vector addition	CH 17_05	166
		- Pythagorean theorem	CH 17_06	167
		- RL circuit impedance	CH 17_07	168

- RL voltage and current	**CH 17_08**	**169**
- RL power	**CH 17_09**	**170**
- RL circuit analysis	**CH 17_10**	**171**
- RC circuit impedance	**CH 17_11**	**172**
- RC voltage and current	**CH 17_12**	**173**
- RC power	**CH 17_13**	**174**
- RC circuit analysis	**CH 17_14**	**175**
- RLC circuit waveforms	**CH 17_15**	**176**
- RLC circuit voltages	**CH 17_16**	**177**
- RLC circuit analysis	**CH 17_17**	**178**

		- RLC power measurement	CH 17_18	179
		- RLC power calculations	CH 17_19	180
		- RLC circuit analysis	CH 17_20	181
		- RLC series resonant circuit	CH 17_21	182
		- RLC resonant circuit analysis	CH 17_22	183
		- RLC resonant frequency calculation	CH 17_23	184
		- RLC resonant frequency Response	CH 17_24	185
Ch. 18	**Resistive, Inductive, Capacitive (RLC) Parallel Circuit**	- RL circuit impedance	CH 18_01	186
		- RL voltage and current	CH 18_02	187

- RL circuit impedance	CH 18_03	188
- RL power	CH 18_04	189
- RL circuit analysis	CH 18_05	190
- RL power calculations	CH 18_06	191
- RC voltage and current	CH 18_07	192
- RC power	CH 18_08	193
- RC circuit analysis	CH 18_09	194
- RC current and power	CH 18_10	195
- RC circuit analysis	CH 18_11	196
- LC current calculations	CH 18_12	197
- LC impedance	CH 18_13	198

		- LC power calculations	CH 18_14	199
		- RLC voltage and current	CH 18_15	200
		- RLC current calculations	CH 18_16	201
		- RLC impedance	CH 18_17	202
		- RLC power measurements	CH 18_18	203
		- RLC parallel resonant circuit	CH 18_19	204
		- Motor load monitoring	CH 18_20	205
		- Inductive load monitoring	CH 18_21	206
		- Motor power factor	CH 18_22	207
		- Motor PF correction	CH 18_23	208
Ch. 19	Transformers	- Turns ratio	CH 19_01	209

- Turns ratio/voltage	CH 19_02	210
- Turns ratio/current	CH 19_03	211
- Circuit analysis	CH 19_04	212
- Polarity test circuit	CH 19_05	213
- Dual voltage transformer	CH 19_06	214
- 3-Phase transformer analysis	CH 19_07	215
- 3-Phase transformer system	CH 19_08	216
- Delta-to-Wye Transformer	CH 19_09	217

SECTION 4 Electrical Installation and Maintenance

	Topic	Lab Assignment	File Mhhe.com/EFT3E	Pg no.
Ch. 20	Circuit Conductors and Wire Sizes	- Line voltage drop	CH 20_01	221
		- Line voltage drop calculation	CH 20_02	222
		- Conductor resistance	CH 20_03	223
		- Conductor length/ resistance	CH 20_04	224
		- Conductor length/ voltage drop	CH 20_04	225
		- Conductor resistance/wire size	CH 20_05	226
Ch. 21	Fuses and Circuit Breakers	- Overloaded circuit	CH 21_01	227
		- Shorted circuit	CH 21_02	228

		- In-circuit fuse test (1)	CH 21_03	229
		- Out-of-circuit fuse test (1)	CH 21_04	230
		- Out-of-circuit fuse test (2)	CH 21_05	231
		- In-circuit fuse test (2)	CH 21_06	232
Ch. 22	Relays	- Relay operation	CH 22_01	233
		- Relay NO and NC contacts	CH 22_02	234
		- Control of high voltage circuit	CH 22_03	235
		- Low current signal control	CH 22_04	236
Ch. 23	Lighting Equipment	- Single-pole switch control	CH 23_01	237
		- Three-way	CH 23_02	238

		switch control		
		- Troubleshooting 3-way switch control	CH 23_03	239
Ch. 24	Electric Motors And Controls	- Across-the-line motor starter	CH 24_01	240
Ch. 25	Electronic Controls	- Diode biasing	CH 25_01	241
		- Light-emitting diode	CH 25_02	242
		- DC operated SCR	CH 25_03	243
		- Triac switching circuit	CH 25_04	244
		- Half-wave rectifier	CH 25_05	246
		- Full-wave rectifier	CH 25_06	247
		- Transistor switching circuit	CH 25_07	248
		- Transistor amplifier circuit	CH 25_08	249

- Digital signal waveform	**CH 25_09**	**250**
- Two-input AND gate	**CH 25_10**	**251**
- Two-input OR gate	**CH 25_11**	**252**
- NOT (Inverter) gate	**CH 25_12**	**253**
- Two-input NAND gate	**CH 25_13**	**254**
- Two-input NOR gate	**CH 25_14**	**255**
-XOR gate	**CH 25_15**	**256**
-Combination gate circuit (1)	**CH 25_16**	**257**
-Combination gate circuit (2)	**CH 25_17**	**258**

SECTION **1**

Fundamentals of Electricity

Chapter 1	Safety	3
Chapter 2	Atoms and Electricity	4
Chapter 3	Sources and Characteristics of Electricity	5
Chapter 4	Electrical Quantities and Ohm's Law	6
Chapter 5	Simple, Series, and Parallel Circuits	13
Chapter 6	Measuring Voltage, Current, and Resistance	30
Chapter 7	Ohm's Law	45
Chapter 8	Resistors	56
Chapter 9	Electricity and Magnetism	66
Chapter 10	Electric Power and Energy	68

CHAPTER 1 Safety

1-1 Open the Multisim circuit file CH01_01.

(a) **Calculate the body current flow** [in amperes (A) and milliamperes (mA)] of an electric shock victim who comes in direct contact with a **120-volt** energy source. Assume a body contact resistance of **1000 ohms.**

(b) **Complete the circuit diagram** with the 120-volt source, ammeter, and 1000-ohm resistor all connected in series.

(c) **Connect the Multisim circuit** according to the diagram. Switch the simulation on and **record the value of the current** flow displayed by the ammeter.

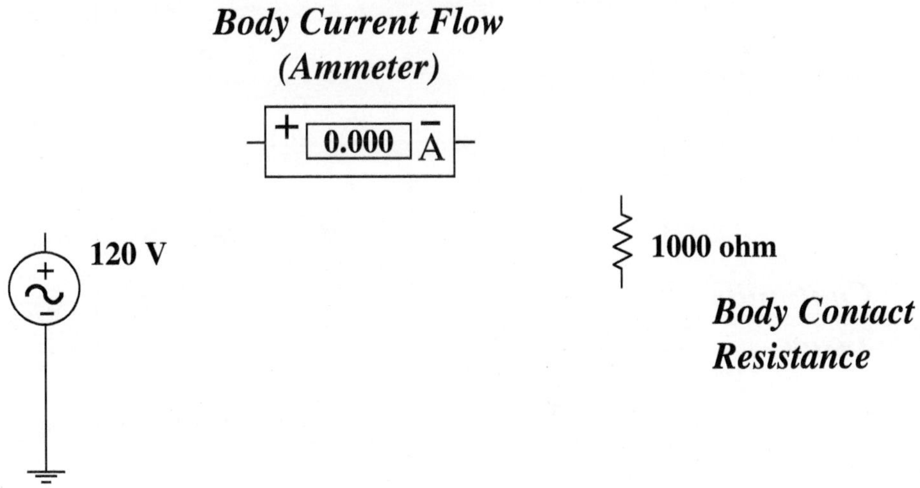

Answers

CHAPTER 2 Atoms and Electricity

2-1 Open the Multisim circuit file CH02_01.

The **continuity tester** shown is to be used to test fuses out-of-circuit for electrical continuity.

(a) **Complete the circuit diagram** with the continuity tester connected in series with fuse F1.

(b) **Connect the Multisim circuit** according to the diagram. Switch the simulation on and **record** whether fuse F1 is **good or bad**. State a **reason** for your answer.

(c) **Repeat** part (b) for fuses **F2, F3, and F4**.

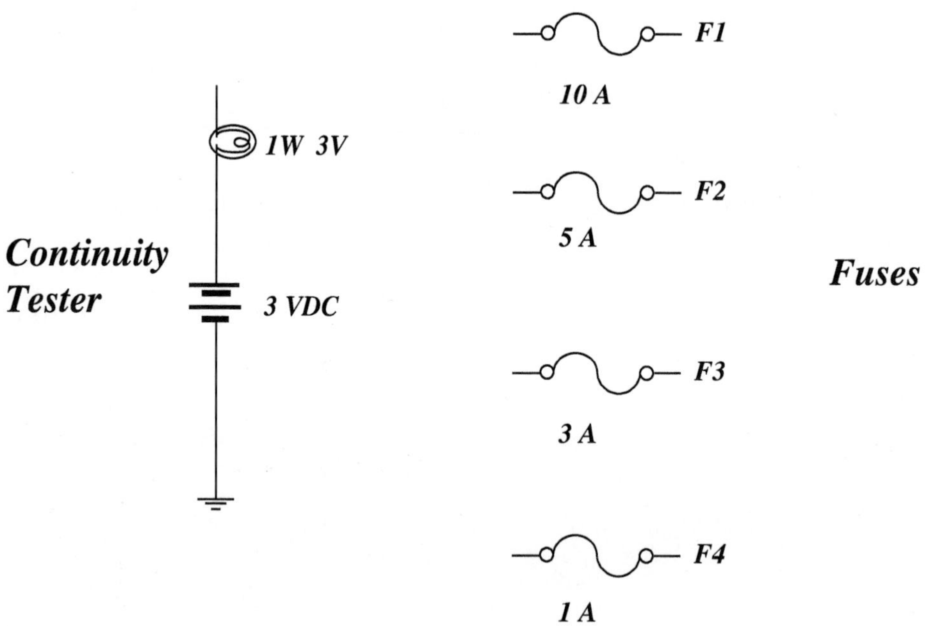

<u>**Answers**</u>

CHAPTER 3 Sources and Characteristics of Electricity

3-1 Open the Multisim circuit file CH03_01.

Current electricity may be classified as **direct current (DC)** or **alternating current (AC)** based upon the voltage source.

(a) Switch the simulation on. Close key A switch, and examine the oscilloscope waveform (double-click on the scope to view the waveform), then **record** whether the voltage source A is **AC or DC**. State a **reason** for your answer.

(b) Open key A switch and close key B switch. Examine the oscilloscope waveform and **record** whether the voltage source B is **AC or DC**. State a **reason** for your answer.

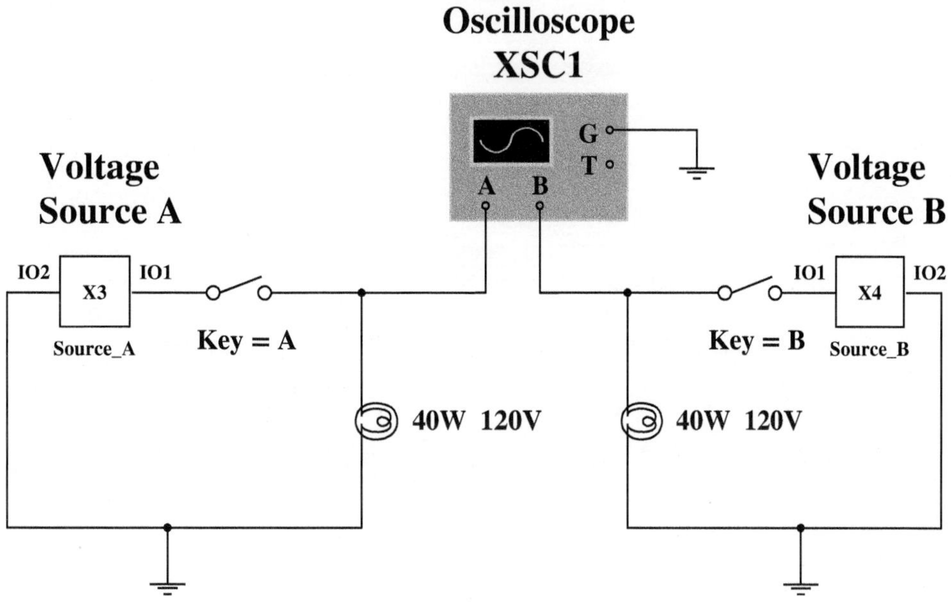

__Answers__

CHAPTER 4 Electrical Quantities and Ohm's Law

4-1 Open the Multisim circuit file CH04_01.

(a) **Complete the circuit diagram** with the ammeter properly connected to measure current.

(b) In what manner (**series or parallel**) is the ammeter connected relative to the lamp load?

(c) **Connect the Multisim circuit** according to the diagram. Switch the simulation on and **record** the amount of **current** of drawn by the lamp.

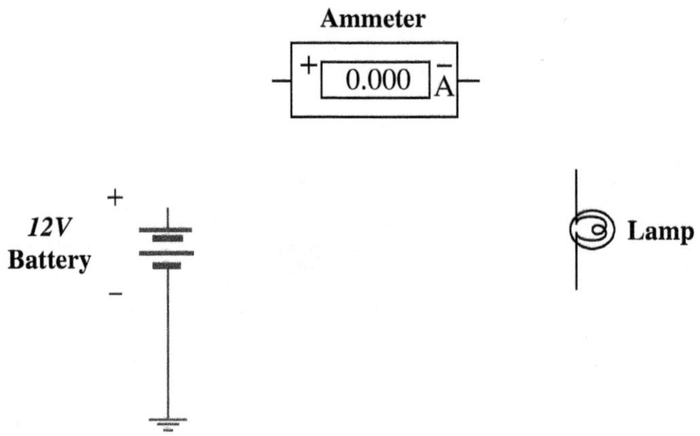

Answers

4-2 Open the Multisim circuit file CH04_02.

(a) **Complete the circuit diagram** by properly connecting voltmeter #1 to measure the battery supply voltage and voltmeter #2 to measure the lamp load voltage.

(b) In what manner (**series or parallel**) are the voltmeters connected?

(c) **Connect the Multisim circuit** according to the diagram. Switch the simulation on and **record** the value of the **supply and load** voltages.

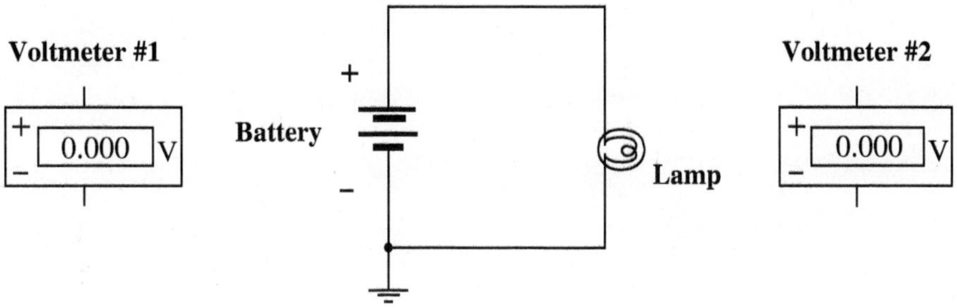

<u>**Answers**</u>

4-3 Open the Multisim circuit file CH04_03.

Switch the simulation on and use the ohmmeter to **measure** and **record the resistance** value of **R1, R2,** and **R3**.

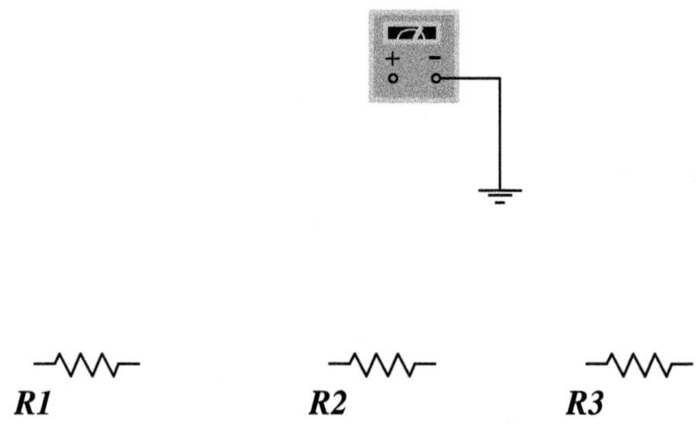

Answers

4-4 Open the Multisim circuit file CH04_04 shown.

(a) Turn the simulation on and **apply 5V** to the 5Ω resistor and **record** the **current** flow.

(b) **Apply 10V** to the 5Ω resistor and **record** the **current** flow.

(c) Which circuit produced the **most current**? **Why**?

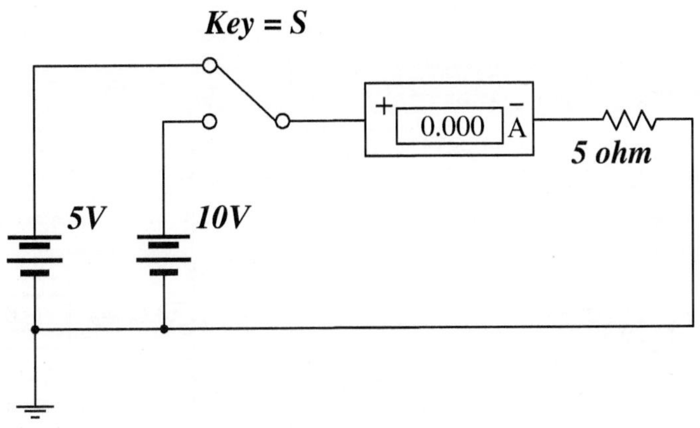

Answers

4-5 Open the Multisim circuit file CH04_05.

(a) If the lamp draws a current of 0.5 A when 120 V is applied to it, **calculate the power** rating of the lamp.

(b) Turn the simulation on and **record** the value of the **voltage, current**, and **power.**

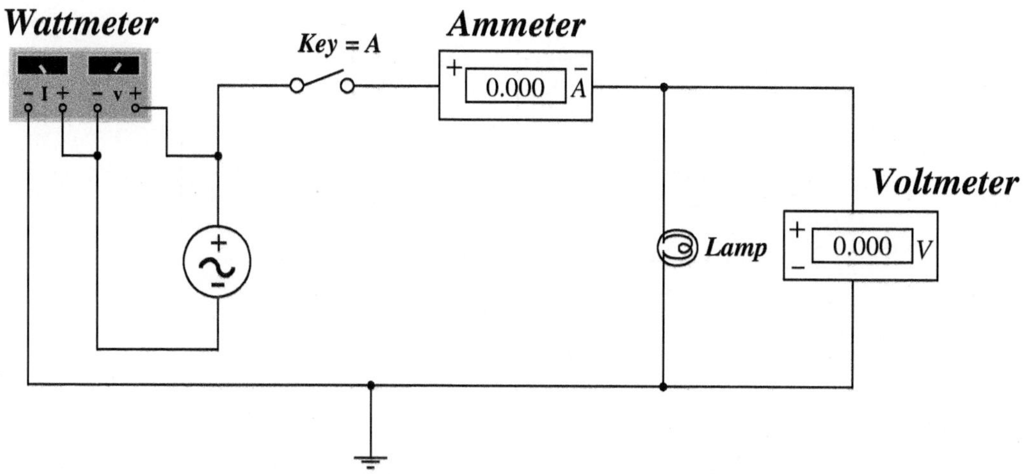

Answers

4-6 Open the Multisim circuit file CH05_06.

(a) Using **Ohm's law,** calculate the value of the **current** flow with **120 volts** applied to the 10 Ω resistor.

(b) Turn the simulation on and **record** the value of the **current.**

(c) Using **Ohm's law,** calculate the value of the **current** flow if the source voltage is decreased to **60 volts.**

(d) Double-click on the source voltage icon and change its value to 60 volts. Turn the simulation on and **record** the value of the **current** with a source voltage of **60 volts** applied.

(e) Using **Ohm's law,** calculate the value of the **current** flow if the source voltage is increased to **240 volts.**

(f) Double-click on the source voltage icon and change its value to 240 volts. Turn the simulation on and **record** the value of the **current** with a source voltage of **240 volts** applied.

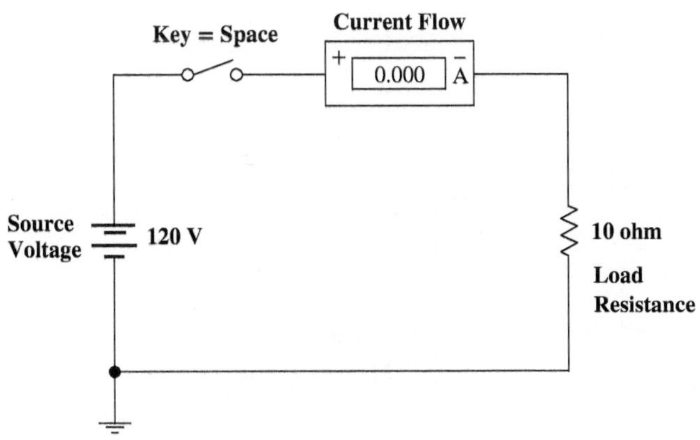

Answers

4-7 Open the Multisim circuit file CH05_07.

(a) Using **Ohm's law,** calculate the value of the **current** flow with **24 volts** applied to the **4 Ω** resistor.

(b) Turn the simulation on and **record** the value of the **current.**

(c) Using **Ohm's law,** calculate the value of the **current** flow if the load resistance is decreased to **2 Ω,** with **24 volts** applied across it.

(d) Double-click on the resistor icon and **change its value to 2 Ω**. Turn the simulation on and **record** the value of the **current** with a load resistance of **2 Ω** connected to 24 volts.

(e) Using **Ohm's law,** calculate the value of the **current** flow if the load resistance is increased to **8 Ω** with **24 volts** applied across it.

(f) Double-click on the resistor icon and **change its value to 8 Ω**. Turn the simulation on and **record** the value of the **current** with a load resistance of **8 Ω** connected to 24 volts.

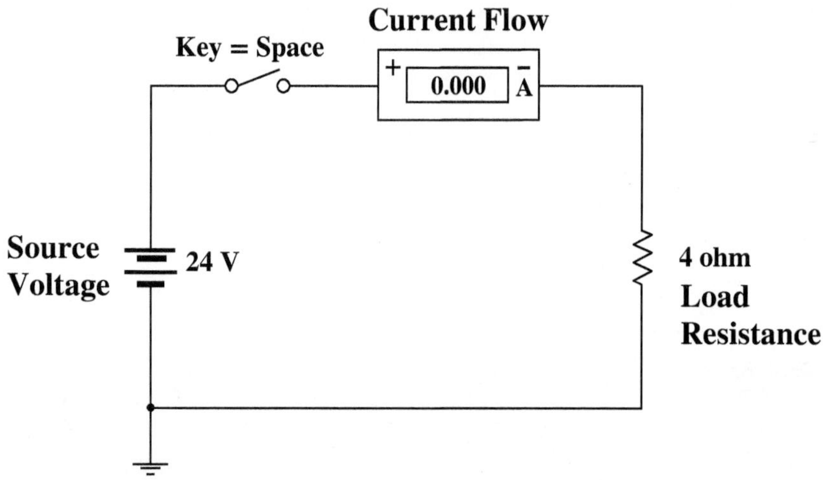

Answers

CHAPTER 5 Simple, Series, and Parallel Circuits

5-1 Open the Multisim circuit file CH05_01.

(a) Turn the simulation on and close switch S. **Record** the value of the **current** flow **into** and **out** of the lamp.

(b) Is the current flow into the lamp **less** than, **greater** than, or the **same** as the current flow out of the lamp?

(c) **Record** the value of the **voltage** across the **lamp**.

(d) Is the voltage across the lamp **less** than, **greater** than, or the **same** as the voltage of the source?

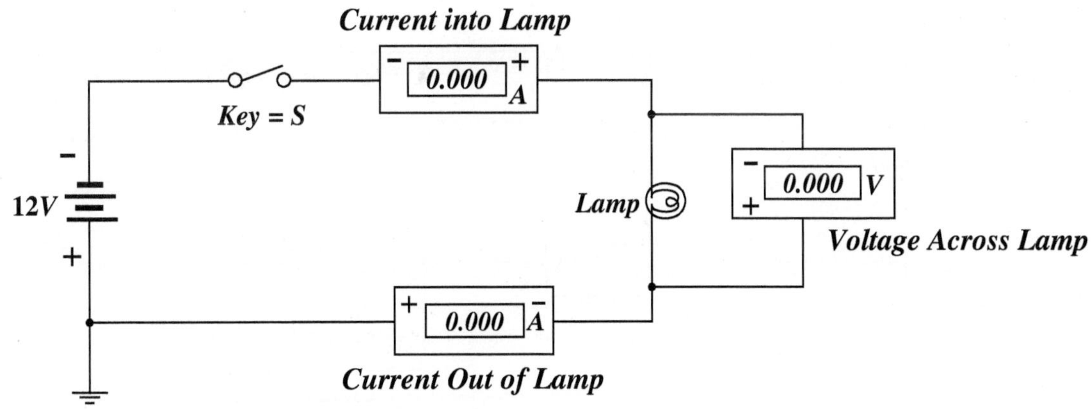

Answers

5-2 Open the Multisim circuit file CH05_02.

(a) What connection of lamps (**series or parallel**) is used?

(b) **Compare** the **voltage** and **power ratings** of the two lamps.

(c) Turn the simulation on and close switch S. **Record** the value of the **voltage** across each lamp.

(d) How do these **voltage values compare** to that of the **source**?

(e) **Record** the value of the **current** at the three points in the circuit.

(f) How do these **current values compare** with one another?

(g) **Replace L2** with the **open lamp** and operate the circuit. Does lamp L1 operate? **Why or why not**?

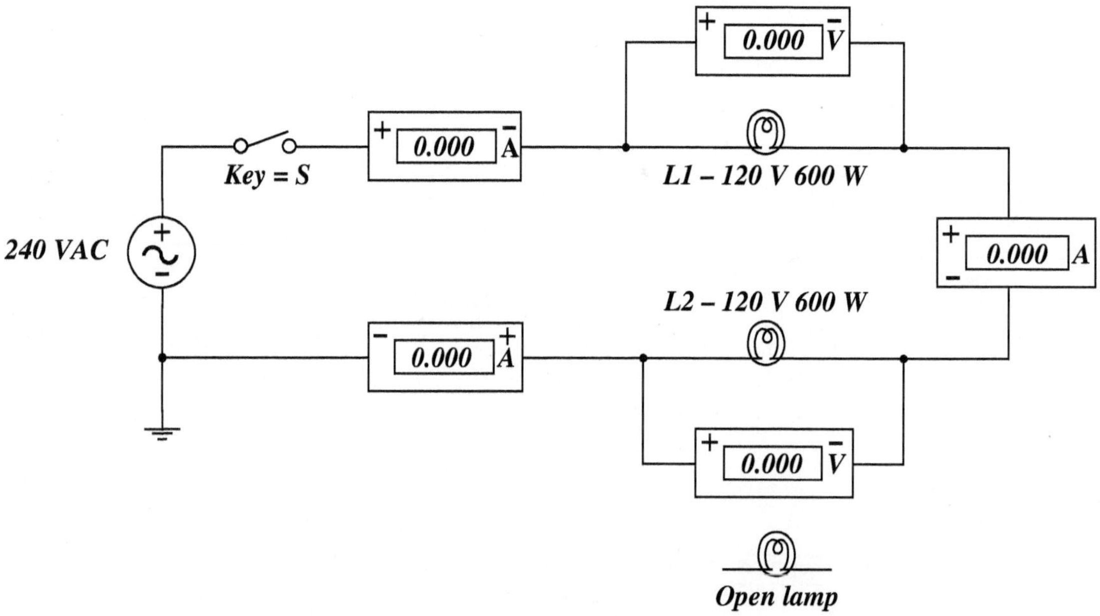

Answers

5-3 Open the Multisim circuit file CH05_03.

(a) What connection of lamps (**series or parallel**) is used?

(b) **Compare** the **voltage** and **power ratings** of the two lamps.

(c) Turn the simulation on and close switch S. **Record** the value of the **voltage** across each lamp.

(d) How do these **voltage values compare** to that of the **source**?

(e) **Record** the value of the **line current, L1 current,** and **L2 current**.

(f) How do these **current values compare** with one another?

(g) **Replace L2** with the **open lamp** and operate the circuit. Does lamp L1 operate? **Why or why not**?

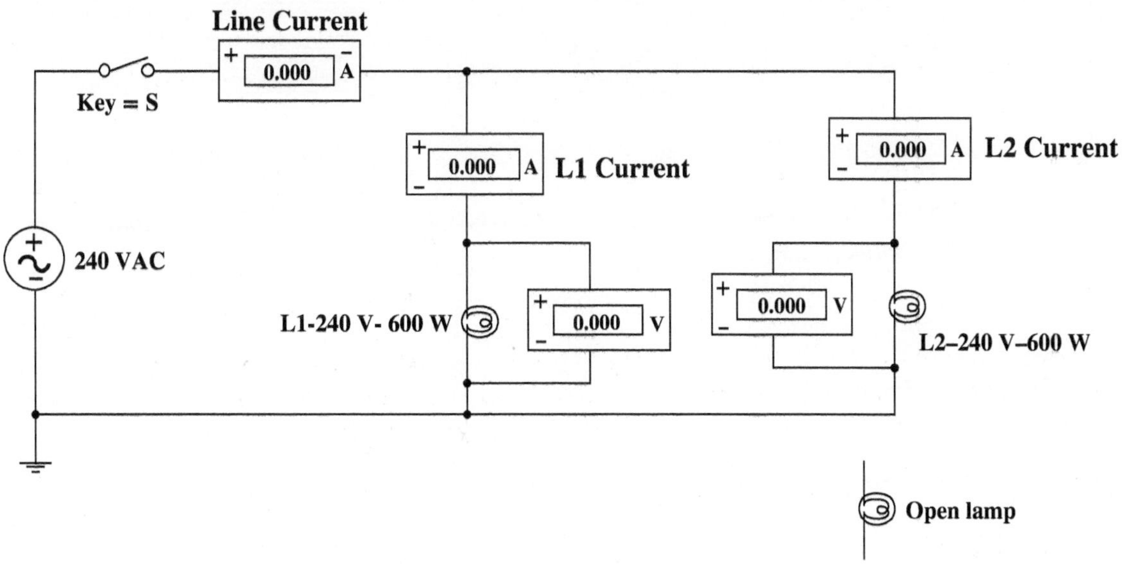

<u>**Answers**</u>

5-4 Open the Multisim circuit file CH05_04.

(a) The 12 V pilot light is to be operated from the 120 V source by connecting the resistor in a manner so as to lower the voltage applied to the lamp. How should the resistor be **connected relative to the lamp** to accomplish this?

(b) **Complete the circuit diagram** showing the correct connection of the resistor and pilot lamp.

(c) **Connect the Multisim circuit** according to the diagram. Operate the simulation and verify that the pilot lamp operates as it should.

(d) **Reconnect** the Multisim circuit so that the resistor and pilot lamp are connected in **parallel** with each other. Operate the simulation and **record what occurs** when the switch is closed.

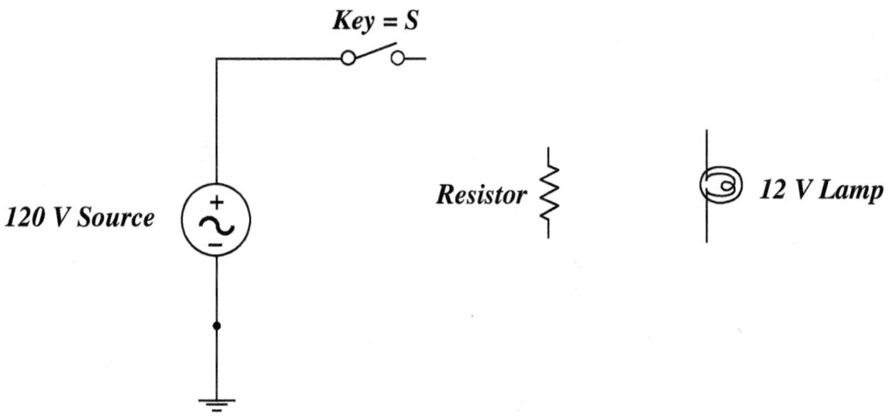

<u>**Answers**</u>

5-5 Open the Multisim circuit file CH05_05.

(a) A control circuit calls for a lamp to come on when **switch A and either switch B or switch C** is closed. In what manner must the **switches be connected** to accomplish this?

(b) **Complete the circuit diagram** showing the correct connection of switches required for this control scenario.

(c) **Connect the Multisim circuit** according to the diagram. Operate the simulation and verify the switching operation.

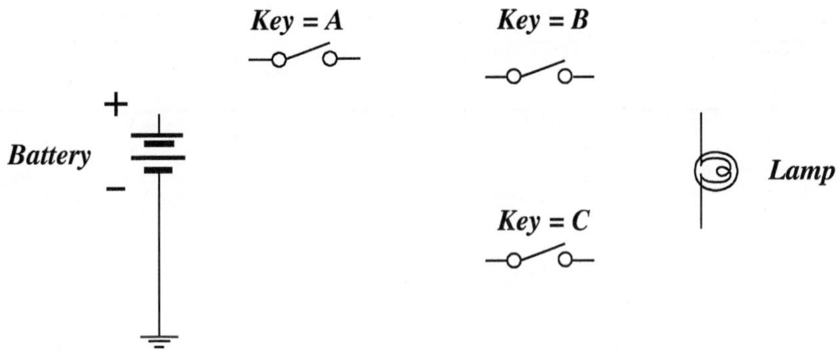

Answers

5-6 Open the Multisim circuit file CH05_06.

(a) What arrangement (**series or parallel**) of lamps is used?

(b) Connect the **spare lamp** into the circuit to determine which lamp is burnt **open**. Record which lamp is opened.

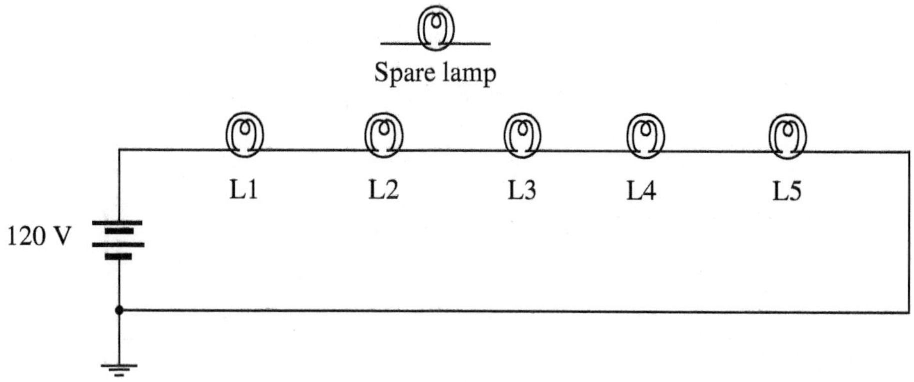

Answers

5-7 Open the Multisim circuit file CH05_07.

(a) Turn the simulation on and with switch **S open, record** the value of the **voltage** across the **switch and each lamp.**

(b) **Close the switch** and **record** the value of the **voltage** across the **switch and each lamp.**

(c) Based on your readings, **why** must the three lamps be **identical in resistance?**

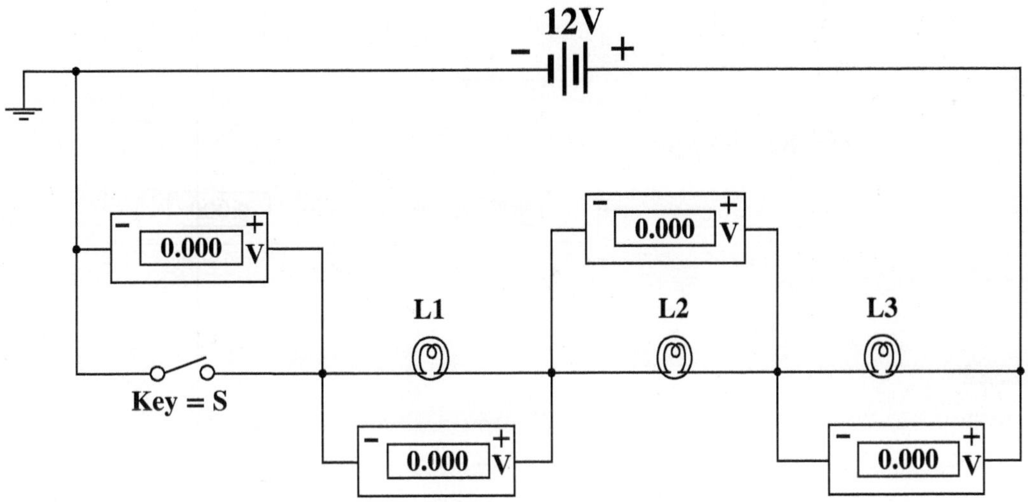

Answers

5-8 Open the Multisim circuit file CH05_08.

(a) Turn the simulation on, close the switch, and **record** the value of the **voltage** across the **resistor and lamp.**

(b) Based on the voltmeter readings, which has the **higher resistance**? **Why?**

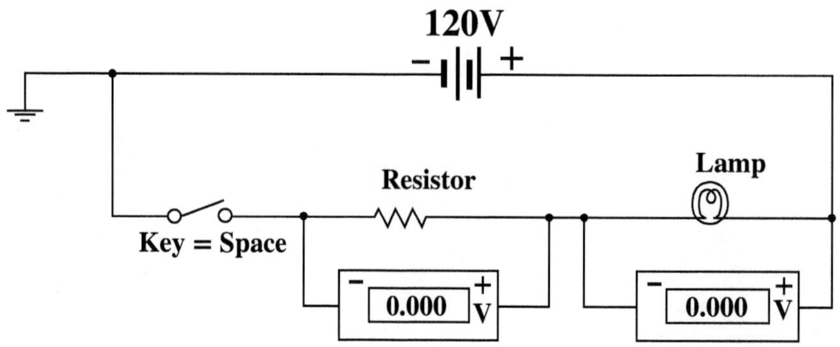

<u>Answers</u>

5-9 Open the Multisim circuit file CH05_09.

(a) Turn the simulation on, close the switches, and **record** the value of the **voltage** across L1 and L2.

(b) Based on the voltmeter readings, what is the value of the **source voltage? Why?**

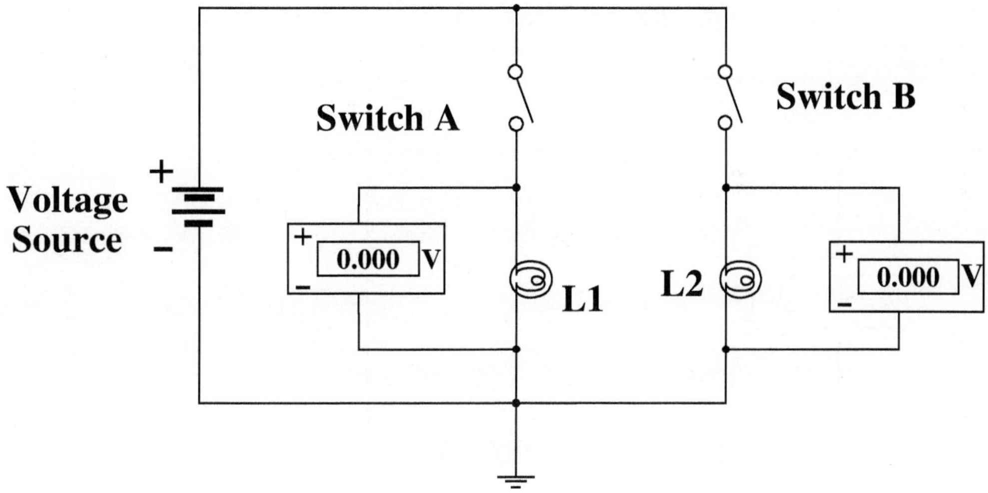

Answers

5-10 Open the Multisim circuit file CH05_10.

Turn the simulation on and operate the circuit.

(a) Record the amount of **current** drawn by the **100 W lamp.**

(b) Record the amount of **current** drawn by the **40 W lamp.**

(c) Which lamp has the **highest resistance**? **Why**?

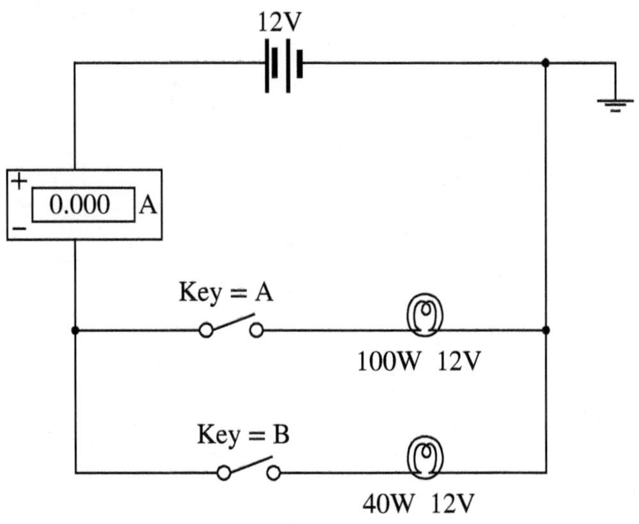

Answers

5-11 Open the Multisim circuit file CH05_11.

(a) **Complete the circuit diagram** with the three lamps connected in **series** with the battery power supply.

(b) **Connect the Multisim circuit** according to the diagram and simulate the normal operation of the circuit.

(c) Simulate the **removal of lamp 3** from the circuit by disconnecting its two wire connections. What **effect,** if any, does this have on the **other lamps**? **Why**?

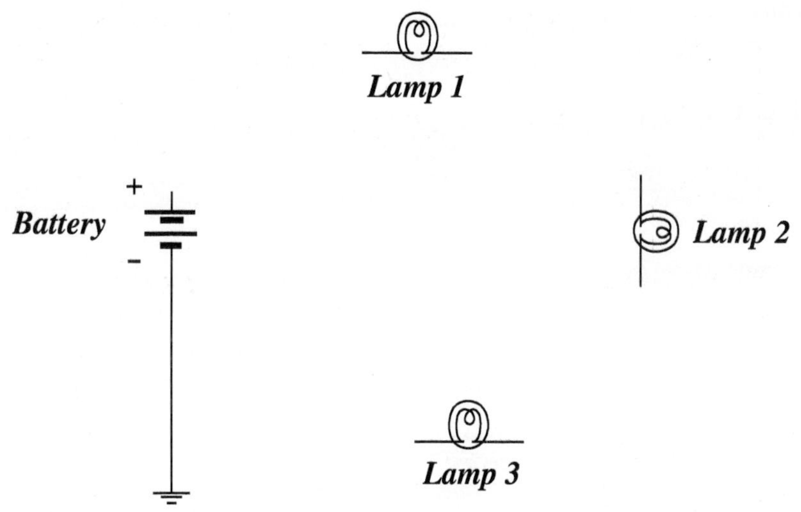

<u>Answers</u>

5-12 Open the Multisim circuit file CH05_12.

(a) **Complete the circuit diagram** with the three lamps connected in **parallel** to the battery power supply.

(b) **Connect the Multisim circuit** according to the diagram and simulate the normal operation of the circuit.

(c) Simulate the **removal of lamp 2** from the circuit by disconnecting its two wire connections. What **effect,** if any, does this have on the **other lamps**? **Why**?

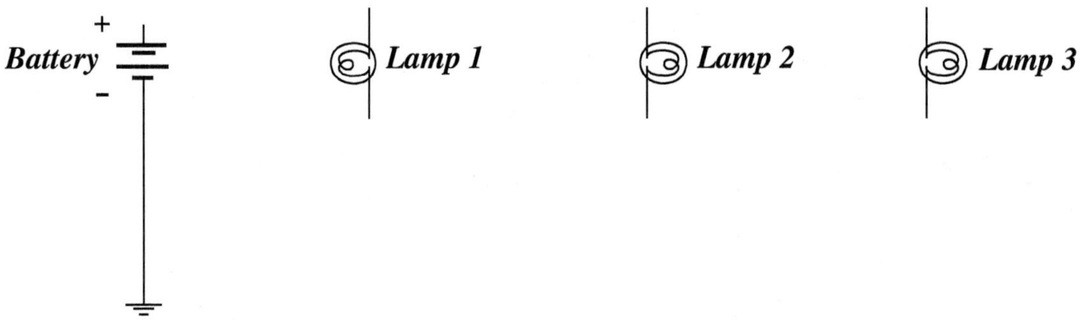

Answers

5-13 Open the Multisim circuit file CH05_13.

(a) **Complete the circuit diagram** with switches A and B connected in **series** to control the lamp.

(b) **Connect the Multisim circuit** according to the diagram and simulate the operation of the circuit.

(c) **Construct a truth table** outlining the manner in which the switches control the lamp.

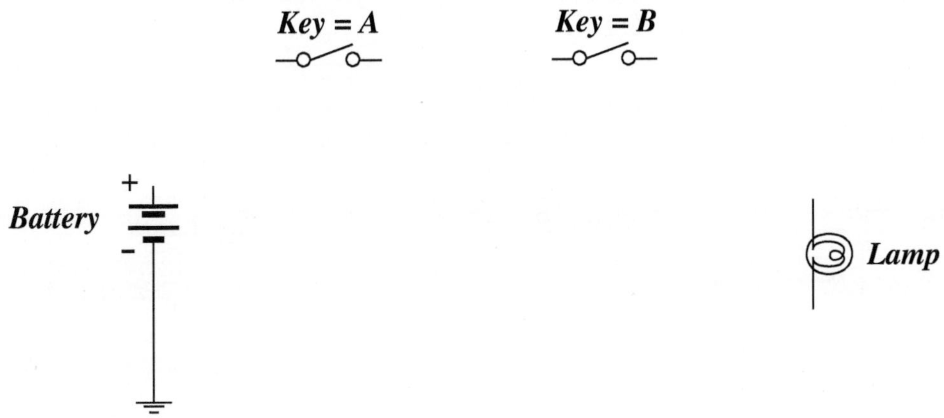

Answers

5-14 Open the Multisim circuit file CH05_14.

(a) **Complete the circuit diagram** with switches A and B connected in **parallel** to control the lamp.

(b) **Connect the Multisim circuit** according to the diagram and simulate the operation of the circuit.

(c) **Construct a truth table** outlining the manner in which the switches control the lamp.

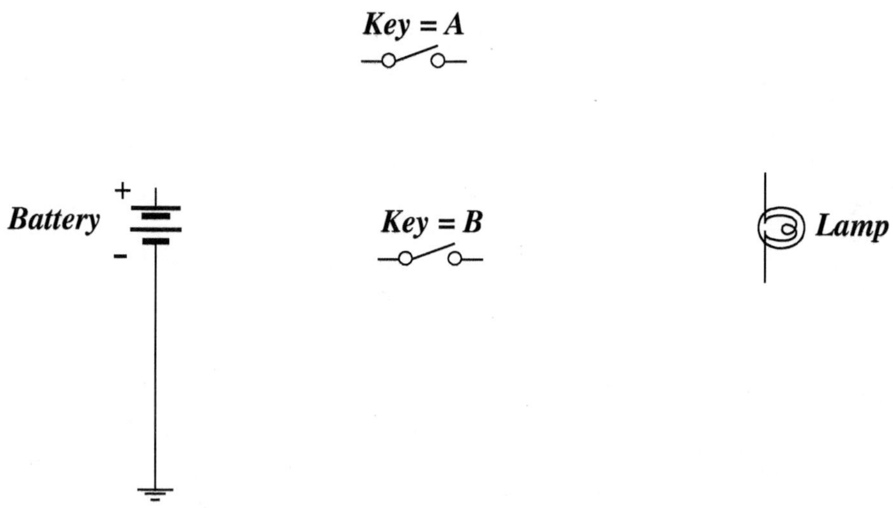

<u>**Answers**</u>

5-15 Open the Multisim circuit file CH05_15.

(a) Turn the simulation on and **operate the circuit**.

(b) **Construct a truth table** outlining the manner in which the switches control the lamp.

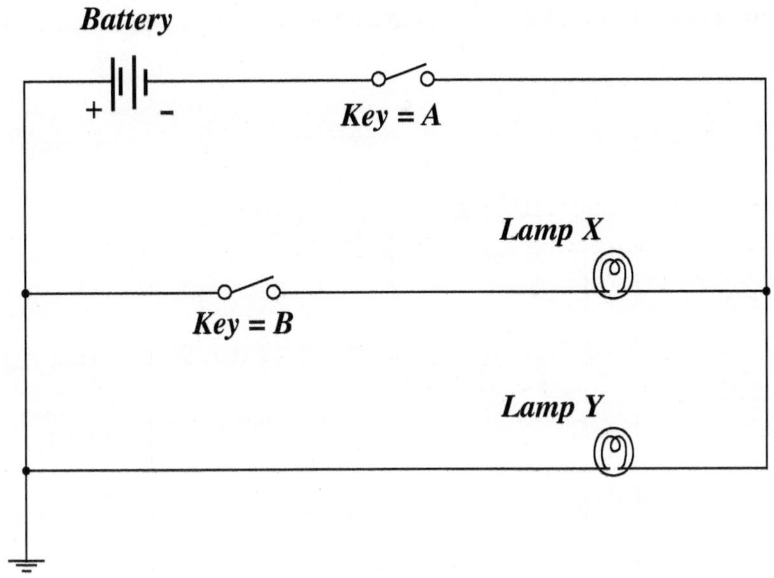

Answers

5-16 Open the Multisim circuit file CH05_16.

(a) Turn the simulation on and **operate the circuit.**

(b) In what way is the wiring of S1 to control L1 **different** from that used for S2 to control L2?

(c) What **difference,** if any, does this have on the way each switch controls each light? **Why?**

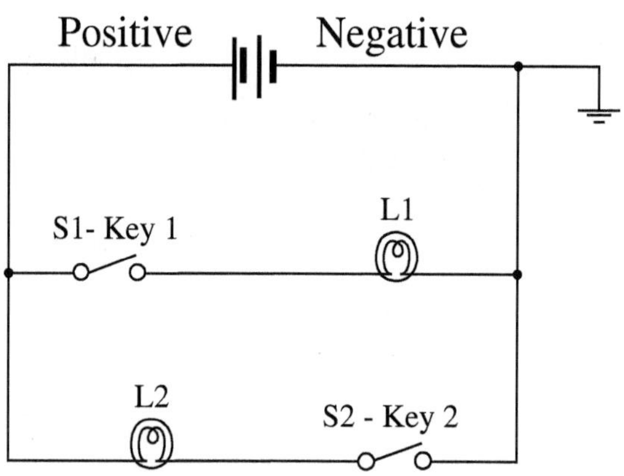

Answers

5-17 Open the Multisim circuit file CH05_17.

(a) Operate the circuit and identify which switch is OK, faulted open, faulted shorted.

(b) Disconnect each switch from the power supply and using the ohmmeter function of the multimeter, measure and record the open and closed resistance of:

(i) switch A (ii) Switch B (iii) Switch C

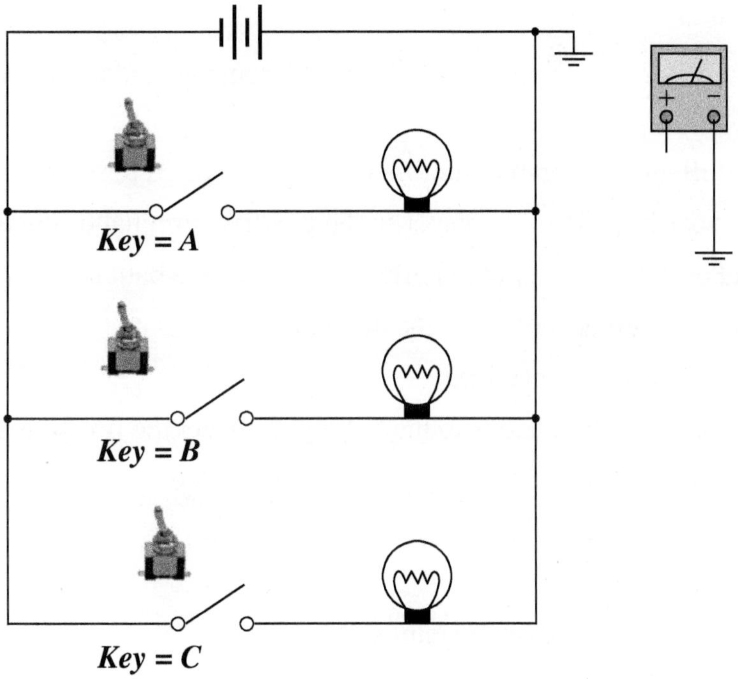

Answers

CHAPTER 6 Measuring Voltage, Current, and Resistance

6-1 Open the Multisim circuit file CH06_01.

Double-click on the multimeter icon to view the multimeter front panel display.

(a) **Complete the circuit diagram** with the multimeter icon connected to measure a DC voltage. Connect the **positive** lead of the meter to the **positive** terminal of the battery and the **negative** lead of the meter to the **negative** terminal of the battery.
(b) **Connect the Multisim circuit** according to the diagram and simulate the operation of the circuit. **Record** the DC voltage **reading.**
(c) **Reverse** the Multisim circuit connections to the battery. Connect the **positive** lead of the meter to the **negative** terminal of the battery and the **negative** lead of the meter to the **positive** terminal of the battery. Operate the simulation and **record** the DC voltage **reading.**
(d) **Explain** the **difference** between the two readings.

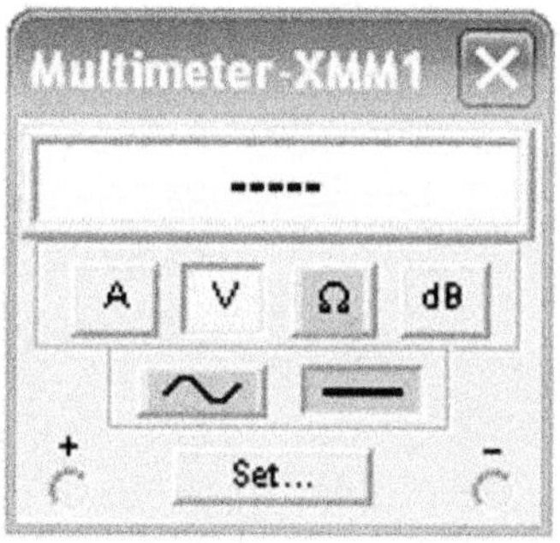

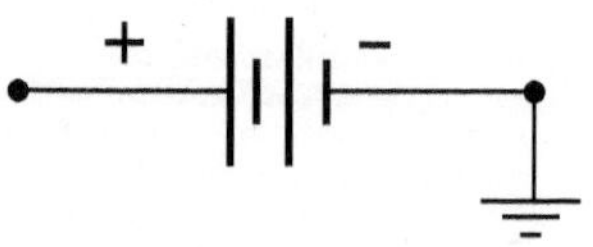

Answers

6-2 Open the Multisim circuit file CH06_02.

The **ohmmeter** function of a multimeter is to be used to check the **continuity** of a single pole switch.

(a) For a **non-faulted** switch, what ohmmeter reading should you expect when the switch is placed in the **on position** and when the switch is placed in the **off position**?
(b) For a switch **faulted open,** what ohmmeter reading should you expect when the switch is placed in the **on position** and when the switch is placed in the **off position**?
(c) For a switch **faulted shorted,** what ohmmeter reading should you expect when the switch is placed in the **on position** and when the switch is placed in the **off position**?
(d) **Connect the Multisim circuit** and record the resistance readings for each switch condition shown. Resistance readings in the **GOhm** range (one GigaOhm equals 1 000 000 000 ohms) are to be considered as **infinite** in value. For v12 simulations, the ohmmeter displays -r- for a very high resistance.

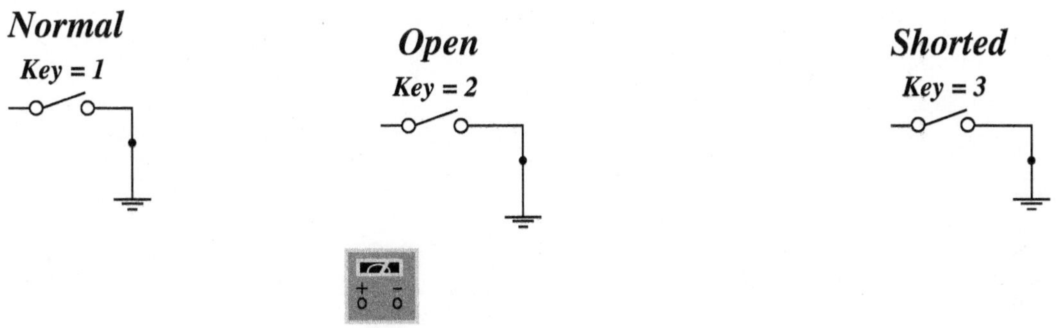

Answers

6-3 Open the Multisim circuit file CH06_03.

(a) Assume that while measuring current in the circuit shown, the two leads of the **ammeter** are accidentally **shorted** together. **What effect** (if any) would this have on the circuit under test? **Why? Verify** your answer by **creating a short across the ammeter** and turning the simulation on.

(b) Assume that while measuring **voltage** in the circuit shown, the two leads of the **voltmeter** are accidentally **shorted** together. **What effect** (if any) would this have on the circuit under test? **Why? Verify** your answer by **creating a short across the voltmeter** and turning the simulation on.

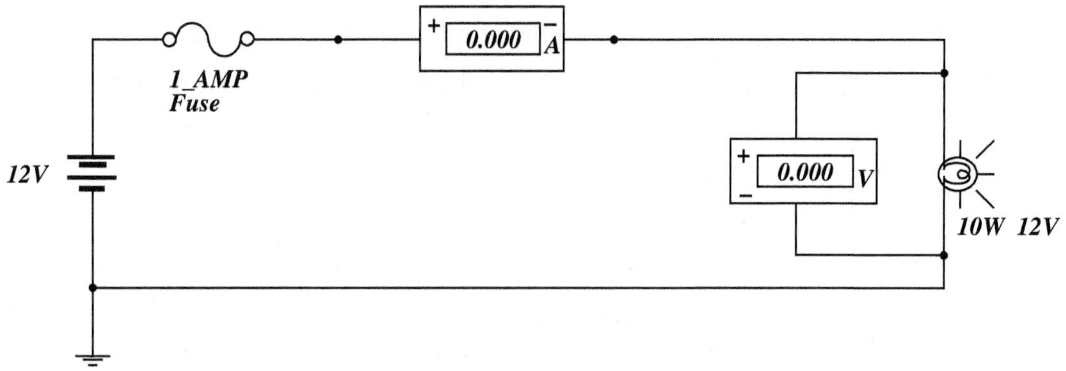

<u>Answers</u>

6-4 Open the Multisim circuit file CH06_04.

Turn the simulation on and operate the circuit.

(a) With the **switch open**, use the multimeter to **record** the voltage between the test points **A, B, C, and the common.**

(b) With the **switch closed**, use the multimeter to **record** the voltage between the test points **A, B, C, and the common.**

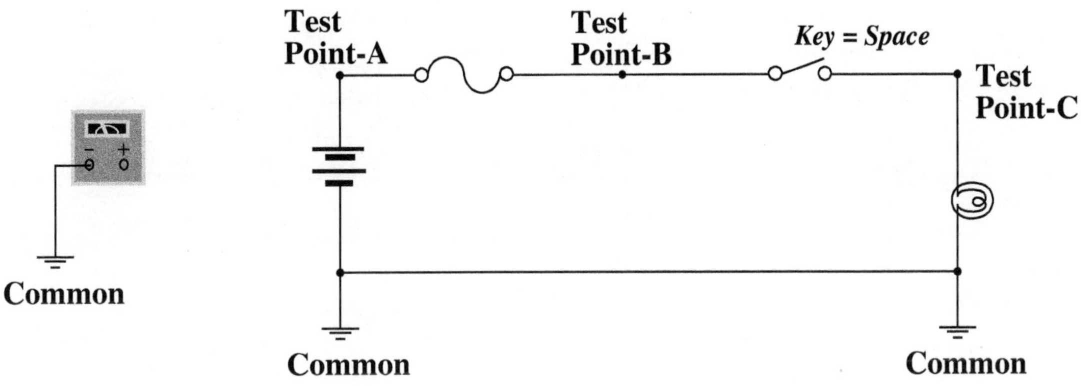

<u>**Answers**</u>

6-5 Open the Multisim circuit file CH06_05.

Load R1 is suspected of having a **lower**-than-normal resistance value.

(a) What **two things** must be done in **preparing the circuit** for a resistance measurement of R1?

(b) Turn the simulation on and use the multimeter to **record** the **resistance value of load R1**.

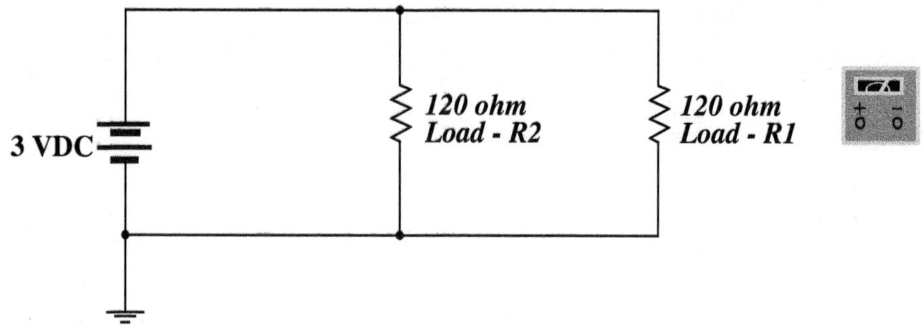

<u>Answers</u>

6-6 Open the Multisim circuit file CH06_06.

Turn the simulation on and operate the circuit.

(a) Use the multimeter to **record** the value of the **voltage** applied to the electronic control module.

(b) Without disconnecting the lead from the circuit, connect the **+ test lead** of the meter to **ground. What happens**? **Why**?

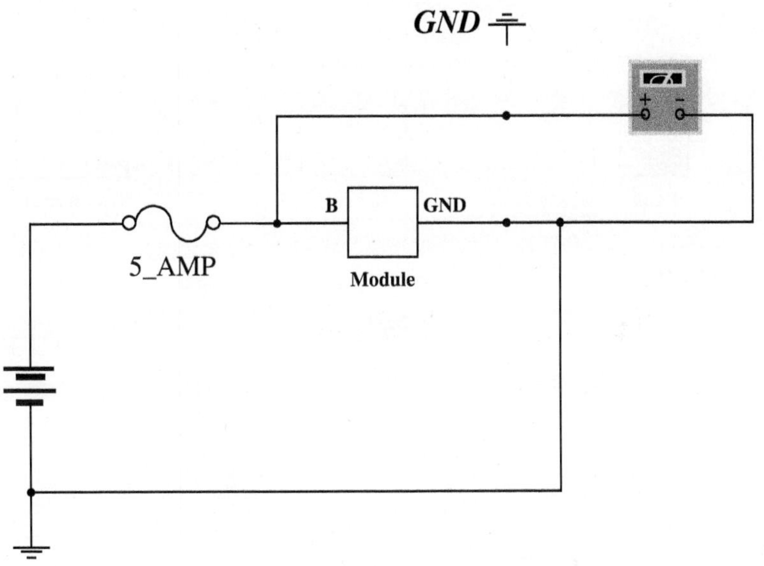

Answers

6-7 Open the Multisim circuit file CH06_07.

Turn the simulation on and operate the circuit.

Connect the multimeter and record the **value** and **type** of voltage (**AC or DC**) obtained from each of the power supply sources (a), (b), (c), and (d).

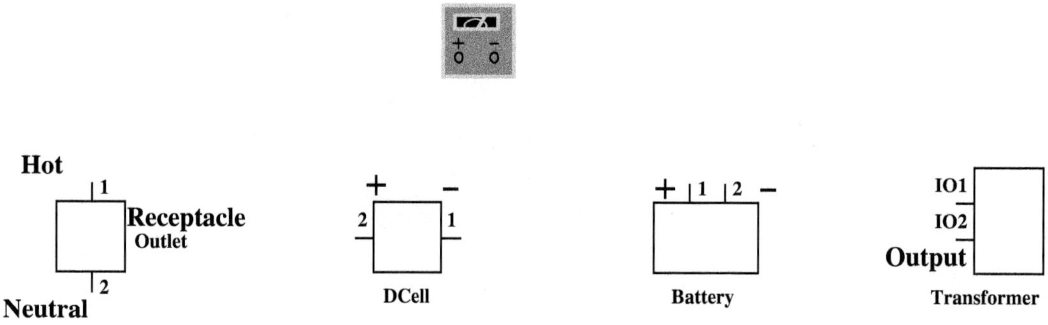

(a) (b) (c) (d)

Answers

6-8 Open the Multisim circuit file CH06_08.

Turn the simulation on and operate the circuit. Connect the multimeter and determine:

(a) The **polarity** of terminals T1 and T2 of DC voltage **source A.**

(b) The **polarity** of terminals T1 and T2 of DC voltage **source B.**

DC Voltage Source A

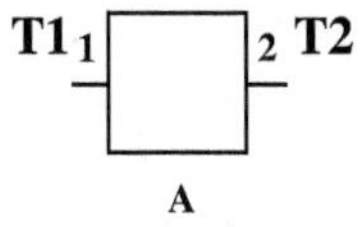

DC Voltage Source B

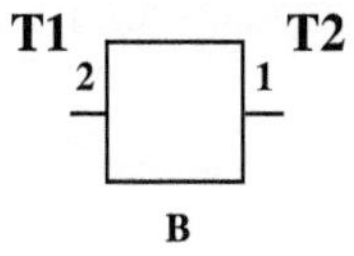

Answers

6-9 Open the Multisim circuit file CH06_09.

Turn the simulation on and operate the series circuit shown.

(a) With the switch closed, use the multimeter to **record** the following:

  i. **battery** voltage
  ii. voltage drop across the **fuse**
  iii. voltage drop across the **switch**
  iv. voltage drop across **lamp L1**
  v. voltage drop across **lamp L2**

(b) Which lamp (L1 or L2) has the **greater** resistance? **Why?**

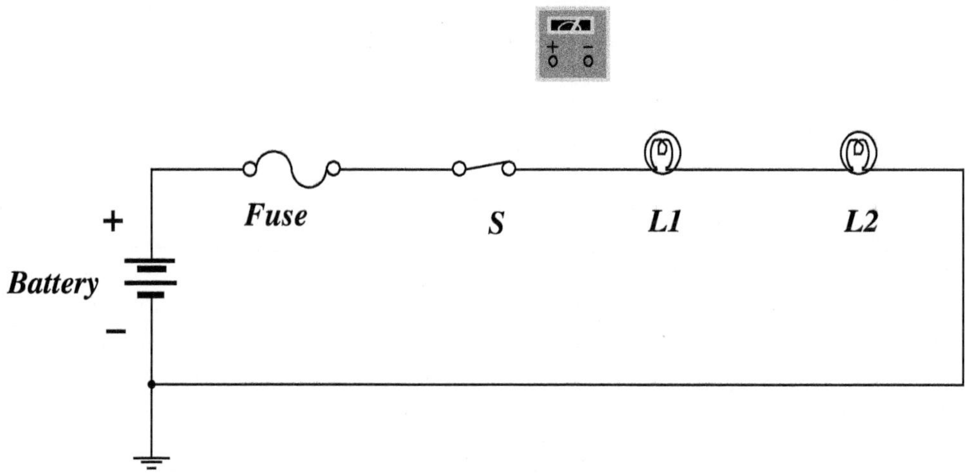

<u>**Answers**</u>

6-10 Open the Multisim circuit file CH06_10.

Use the multimeter to record the amount of current that flows when **240 VAC** is **applied in turn** to each of the following loads:

(a) 4 ohm
(b) 10 ohm
(c) 24 ohm
(d) 100 ohm

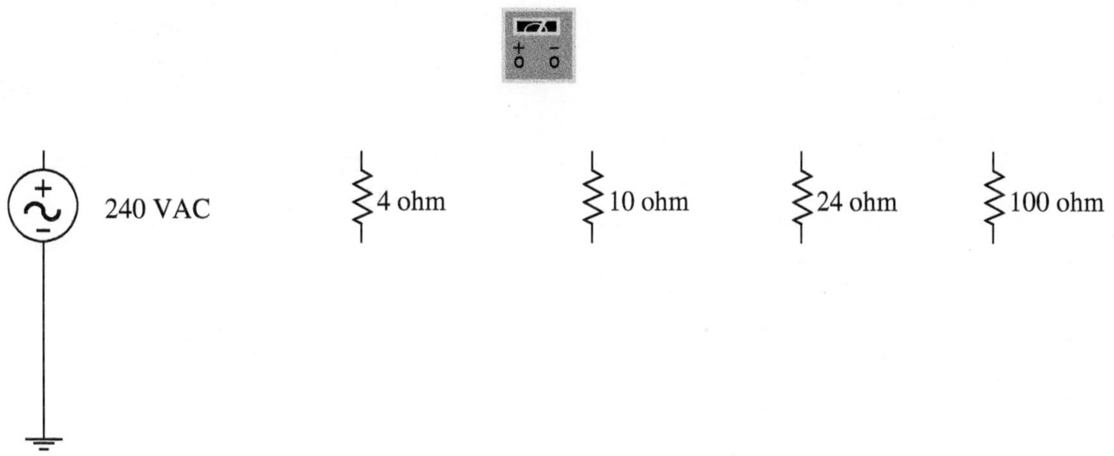

Answers

6-11 Open the Multisim circuit file CH06_11.

Use the multimeter to **record the resistance** of each of the following loads:

(a) R1

(b) R2

(c) R3

(d) R4

(e) R5

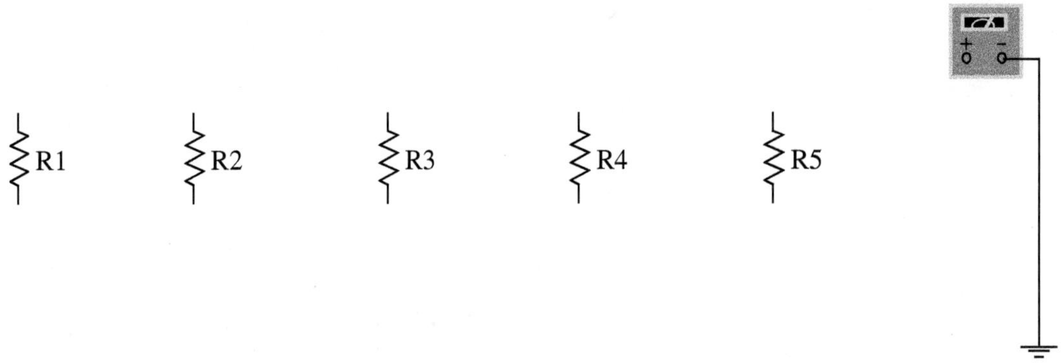

Answers

6-12 Open the Multisim circuit file CH06_12.

Use the multimeter to perform a **continuity test** of each component and classify each as being **OK** or **defective**.

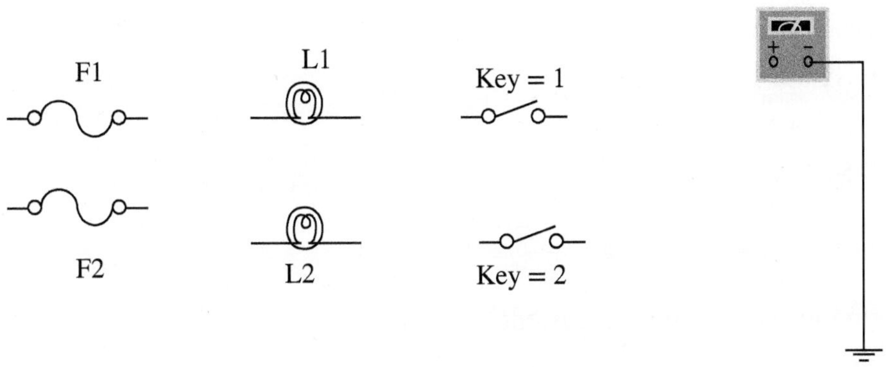

Answers

6-13 Open the Multisim circuit file CH06_13.

Turn the simulation on and operate the parallel circuit shown. **Record** the **total current** for each of the following conditions:

(a) S1 closed – S2 & S3 open
(b) S2 closed – S1 & S3 open
(c) S3 closed – S1 & S2 open
(d) S1 & S2 closed – S3 open
(e) S1 & S3 closed – S2 open
(f) S2 & S3 closed – S1 open
(g) S1, S2, and S3 closed
(h) Which combination results in the most current?

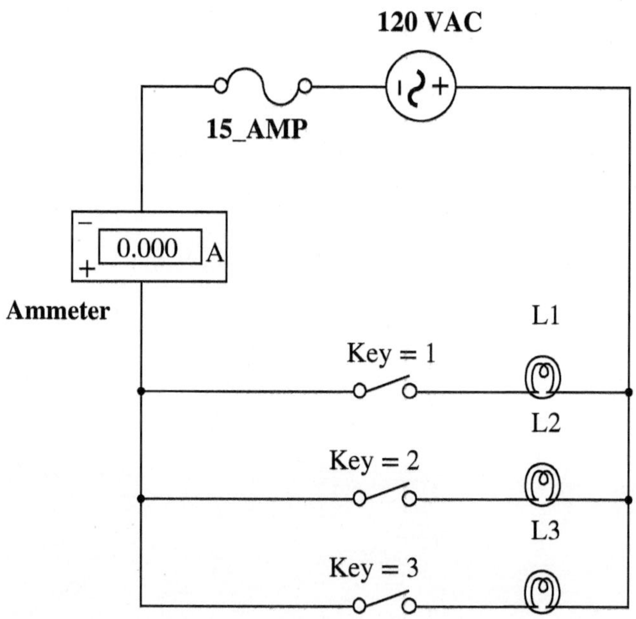

<u>Answers</u>

6-14 Open the Multisim circuit file CH06_14.

(a) What connection of lamps (**series** or **parallel**) is used?

(b) Connect the ammeter and **record** the amount of **current** flow.

(c) Connect the **voltmeter** and **record** the amount of voltage drop across each lamp.

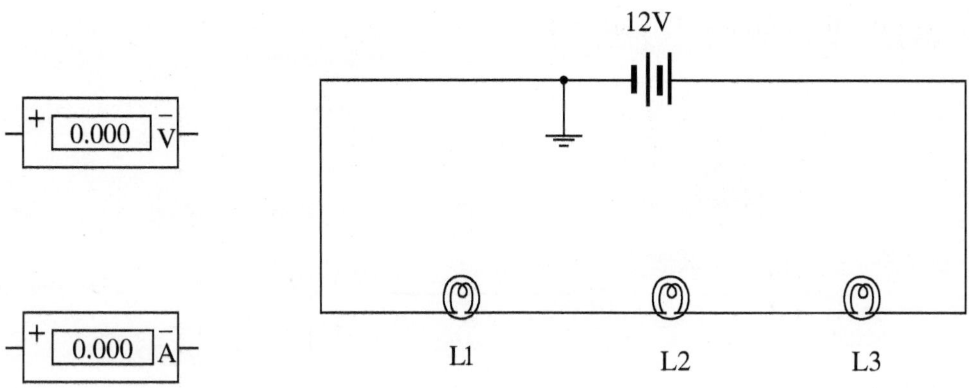

Answers

6-15 Open the Multisim circuit file CH06_15.

Turn the simulation on and operate the circuit. Connect the multimeter and **record** the value of:

(a) The DC **source voltage**

(b) The **voltage** drop across **R1, L1,** and **L2.**

(c) The **current** flow through **R1, L1,** and **L2.**

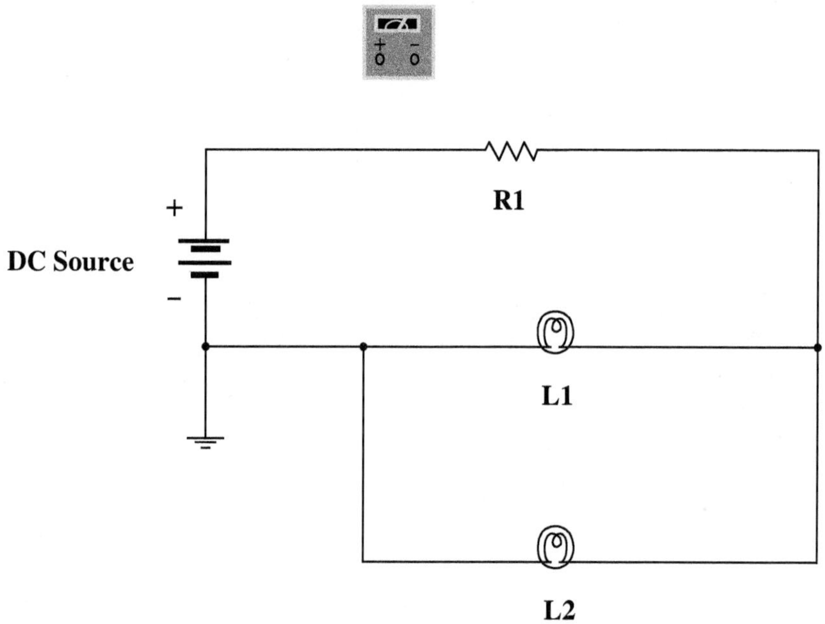

Answers

CHAPTER 7 Ohm's Law

7-1 Open the Multisim circuit file CH07_01.

(a) For **each circuit** drawn [(a) through (e)], **calculate** the expected **current** flow according to **Ohm's law.**

(b) **Measure** and **record** the actual amount of **current flow** for each circuit.

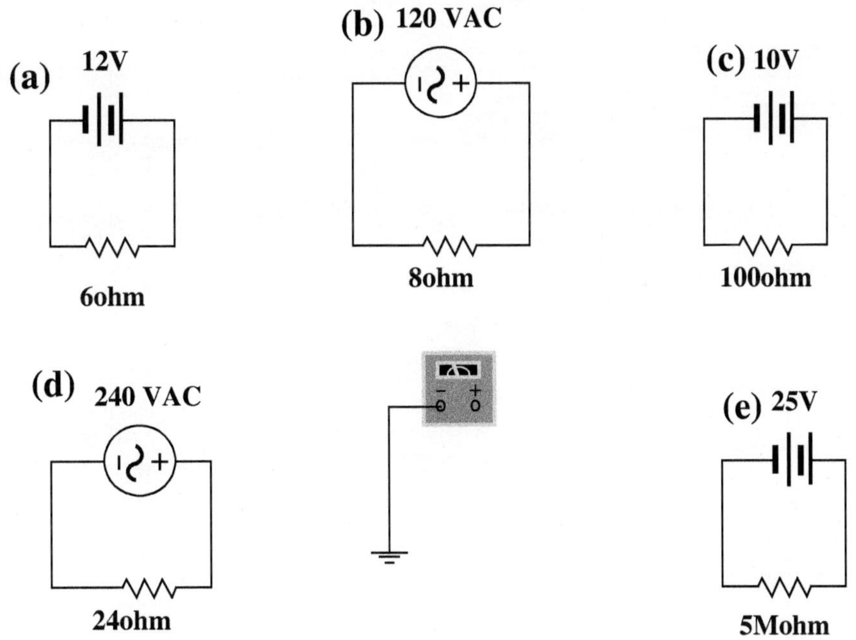

<u>**Answers**</u>

7-2 Open the Multisim circuit file CH07_02.

(a) Connect **each resistor**, in turn, to the **100-volt** source. Turn the simulation on and **record** the amount of circuit **current** flow for each resistance value.

(b) According to your results, explain the effect that the changes in circuit resistance have on the current flow.

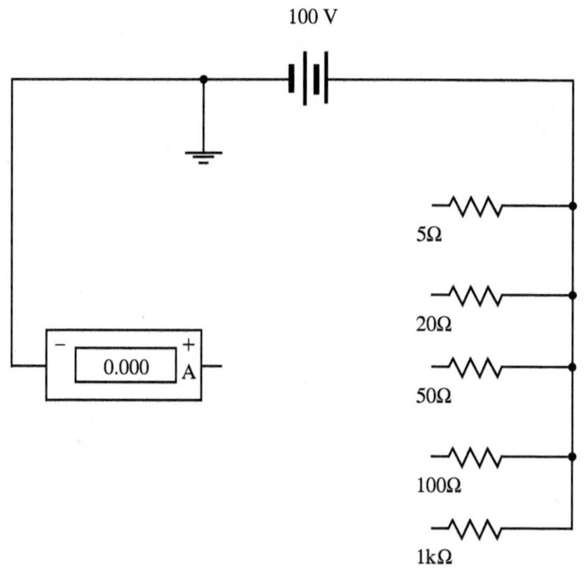

Answers

7-3 Open the Multisim circuit file CH07_03.

(a) For **each circuit** shown, **calculate** the expected applied **voltage** according to **Ohm's law.**

(b) **Measure** and **record** the actual amount of voltage applied to each circuit.

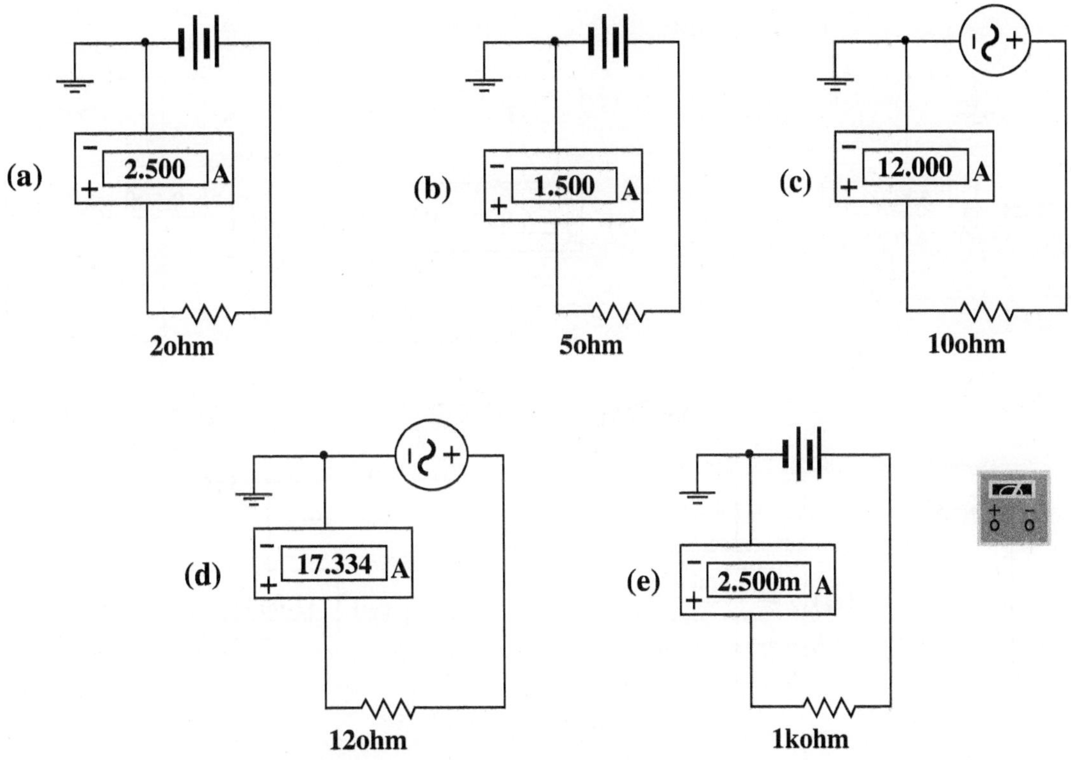

<u>**Answers**</u>

7-4 Open the Multisim circuit file CH07_04.

(a) For **each circuit** shown, **calculate** the expected load **resistance** according to **Ohm's** law.

(b) **Measure** and **record** the actual resistance of the load.

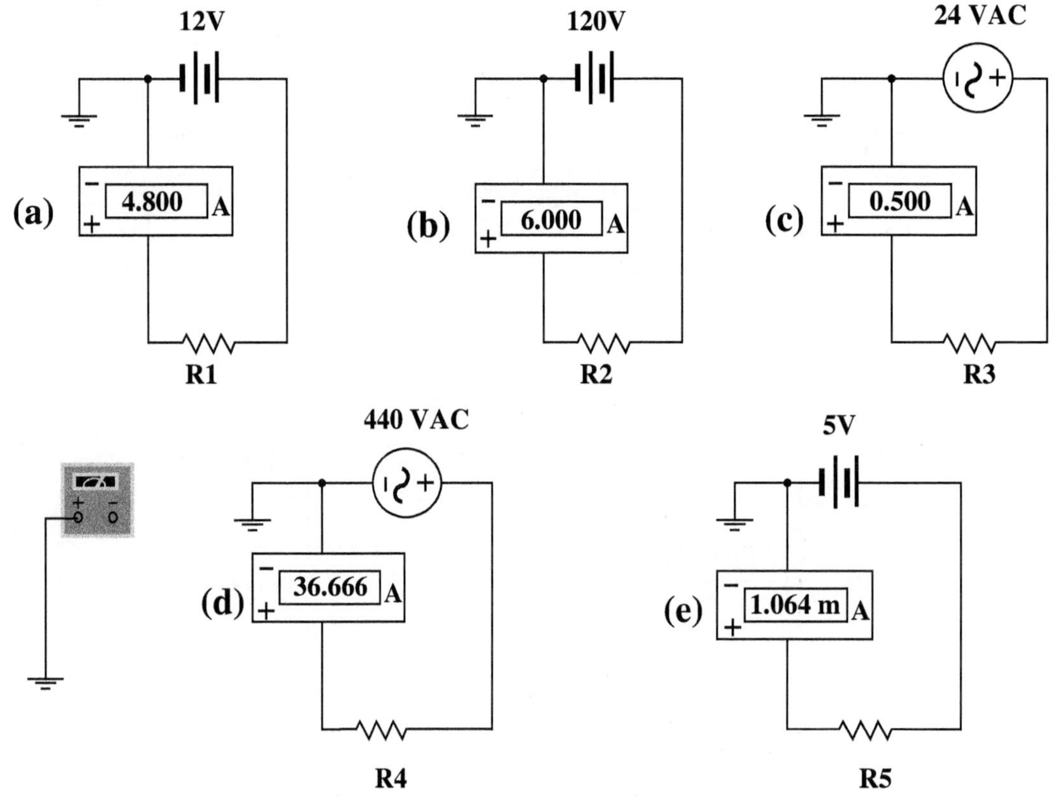

Answers

7-5 Open the Multisim circuit file CH07_05.

(a) **Change** the supply voltage (**in 1-V steps**) from 0 to 10 volts. **Record** the amount of **current** for each voltage step.

(b) Based on your results, explain the **effect on current** of changes in **voltage**.

(c) Plot a graph of **current versus voltage**. With reference to this graph, explain why the resistor would be classified as a linear device.

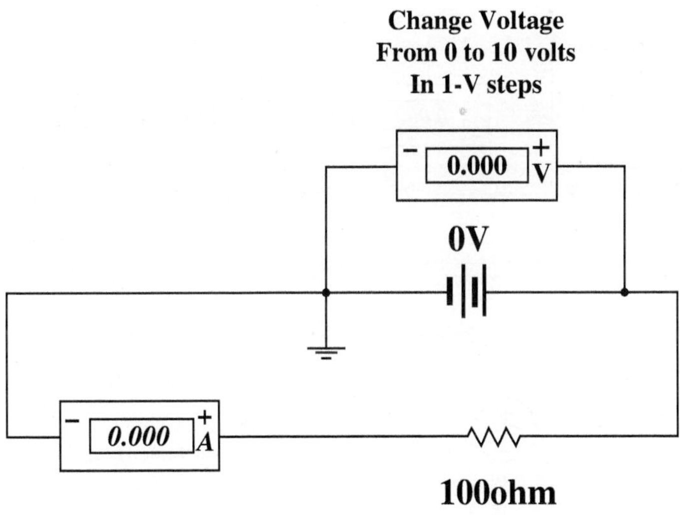

<u>**Answers**</u>

7-6 Open the Multisim circuit file CH07_06.

Turn the simulation on and **observe the meter readings** for each of the three circuits. Using the **power formulas,** determine the:

(a) wattage of the **lamp**

(b) power dissipated by the **resistor**

(c) power rating of the portable **heater**

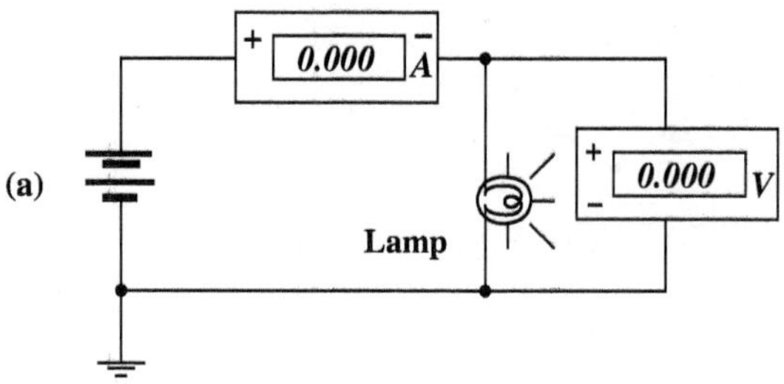

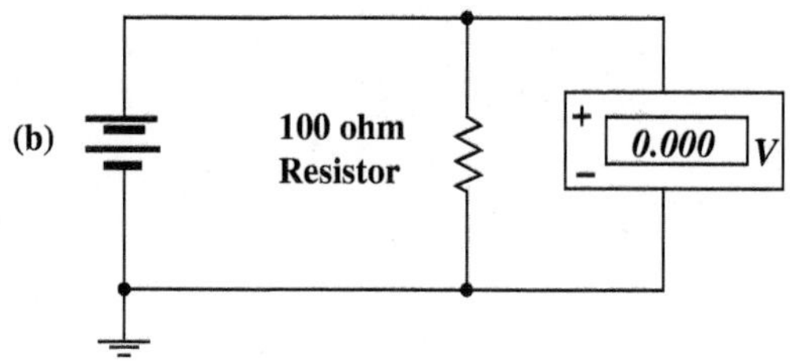

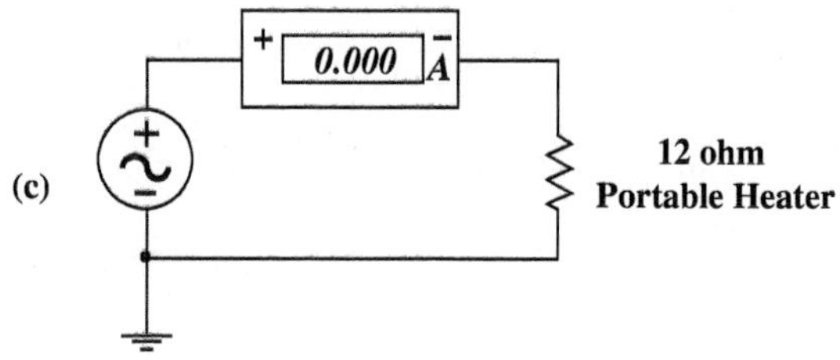

<u>Answers</u>

7-7 Open the Multisim circuit file CH07_07.

Turn the simulation on and observe the **ammeter reading**.

(a) **Calculate** the value of the **voltage drop** across the resistor.

(b) Use the multimeter to **record value** of the voltage drop across the resistor.

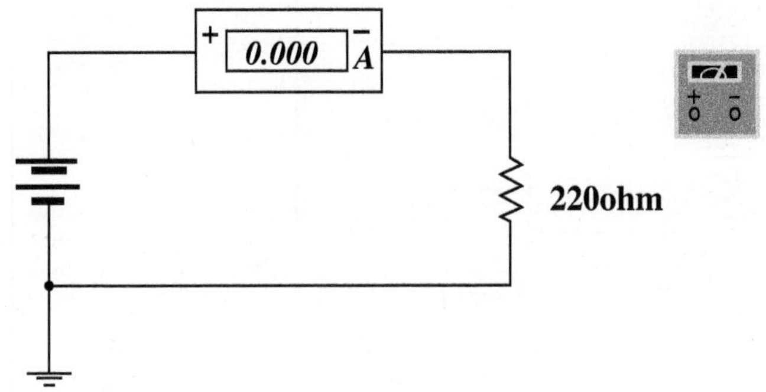

Answers

7-8 Open the Multisim circuit file CH07_08.

(a) **Determine** the value of the **current** flow through module.

(b) Turn the simulation on and **measure** the value of the **current** flow.

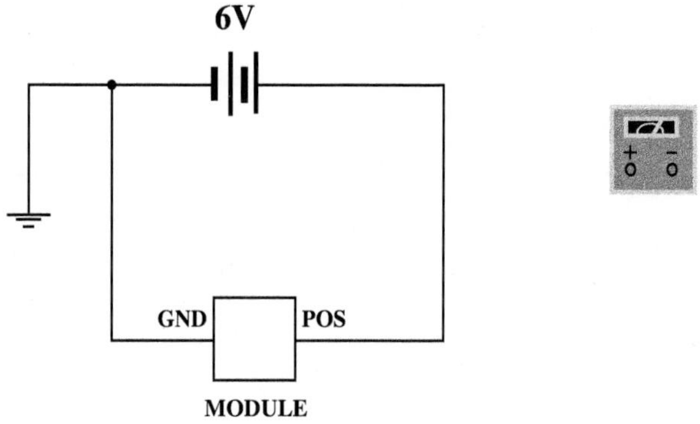

Resistance = 50,000 ohms

Answers

7-9 Open the Multisim circuit file CH07_09.

Turn the simulation on and **observe the meter readings**.

(a) **Calculate** the value of **resistor R.**

(b) Turn the simulation on and **measure** the **resistance of R.**

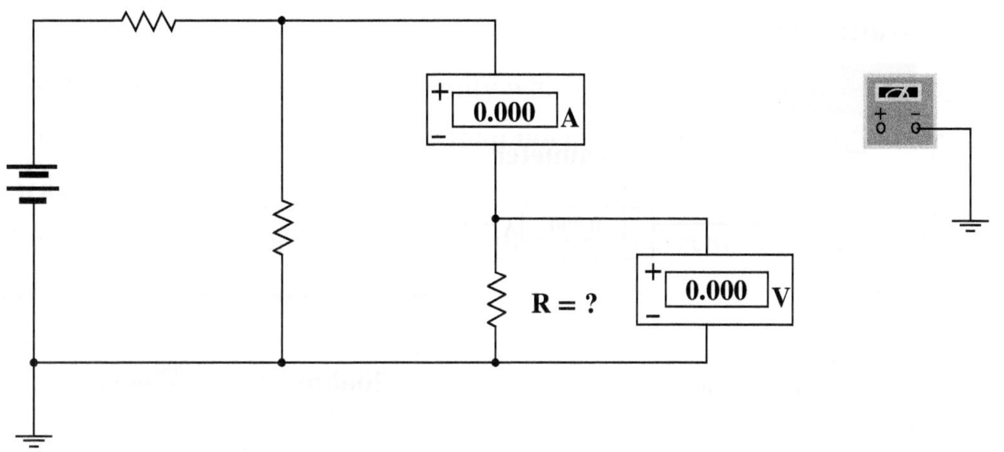

Answers

CHAPTER 8 Resistors

8-1 Open the Multisim circuit file CH8_01.

The **variable resistor** shown is designed to provide for **continuous adjustment of resistance.**

Turn the simulation on and operate the variable resistor. Record the resistance reading for each of the following positions:

(a) 0 %
(b) 25 %
(c) 50 %
(d) 75 %
(e) 100 %

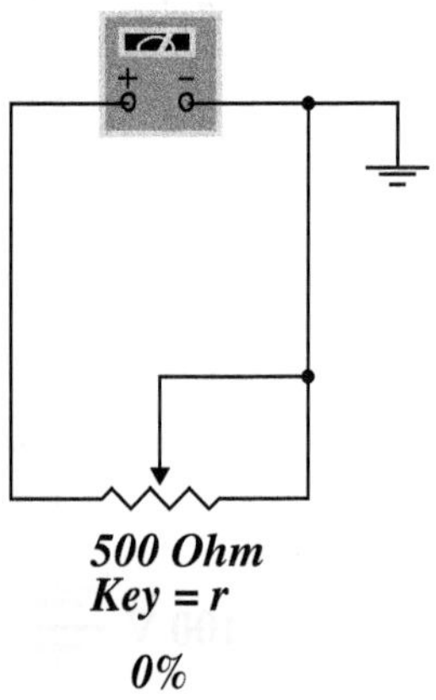

Answers

8-2 Open the Multisim circuit file CH8_02.

The circuit shown uses a **rheostat** to **control current** by varying the resistance in a circuit.

Turn the simulation on and operate the circuit. **Record** the **current** for the following:

(a) 0 % (zero Ω) setting of the rheostat
(b) 25 % (2.5 Ω) setting of the rheostat
(c) 50 % (5 Ω) setting of the rheostat
(d) 75 % (7.5Ω) setting of the rheostat
(e) 100 % (10 Ω) setting of the rheostat

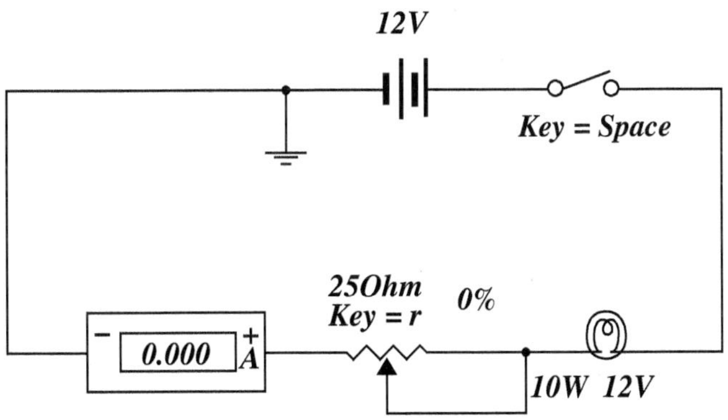

Answers

8-3 Open the Multisim circuit file CH8_03.

The circuit shown uses a **potentiometer** to **adjust** the level of a DC **voltage.**

Turn the simulation on and operate the circuit. **Record** the **voltage** obtained for each of the following settings of the potentiometer:

(a) 0 %
(b) 20 %
(c) 40 %
(d) 60 %
(e) 80 %
(f) 100 %

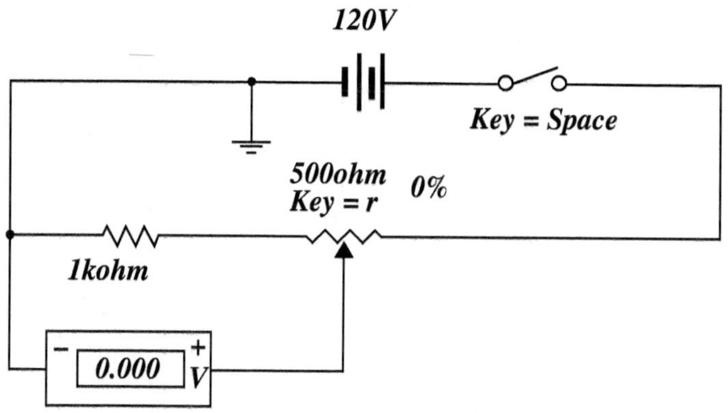

Answers

8-4 Open the Multisim circuit file CH8_04.

(a) **Calculate** the **total resistance** of the series connected resistor circuit shown.

(b) Turn the simulation on and **record the measured value** of the **total resistance** of the circuit.

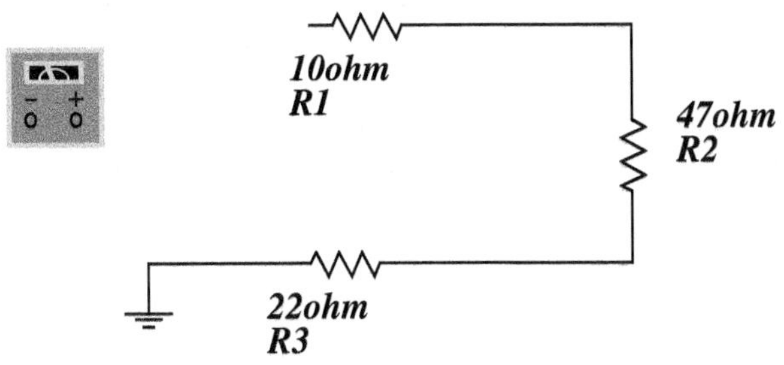

Answers

8-5 Open the Multisim circuit file CH8_05.

(a) **Calculate** the voltages **E1, E2, and E3** for the **voltage-divider** network shown.

(b) Turn the simulation on. **Measure** and **record** the **five** different values of DC **voltage** available from this network.

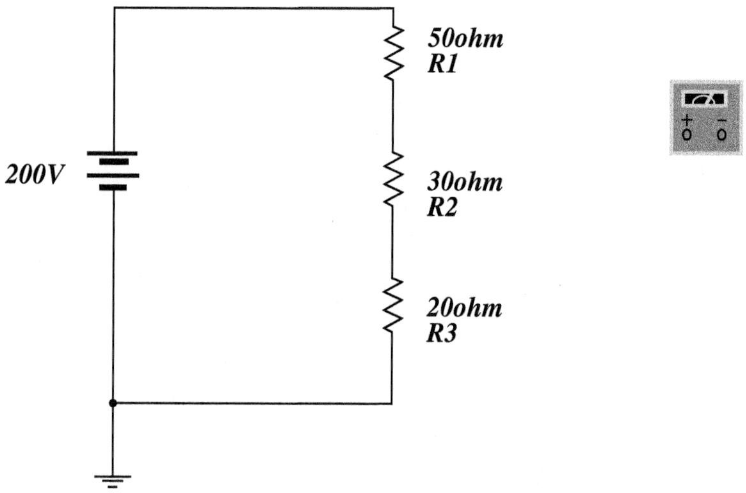

Answers

8-6 Open the Multisim circuit file CH8_06.

(a) **Calculate** the **total resistance** for each circuit [**(a) through (f)**].

(b) Turn the simulation on. **Measure** and **record** the **total resistance** of each circuit.

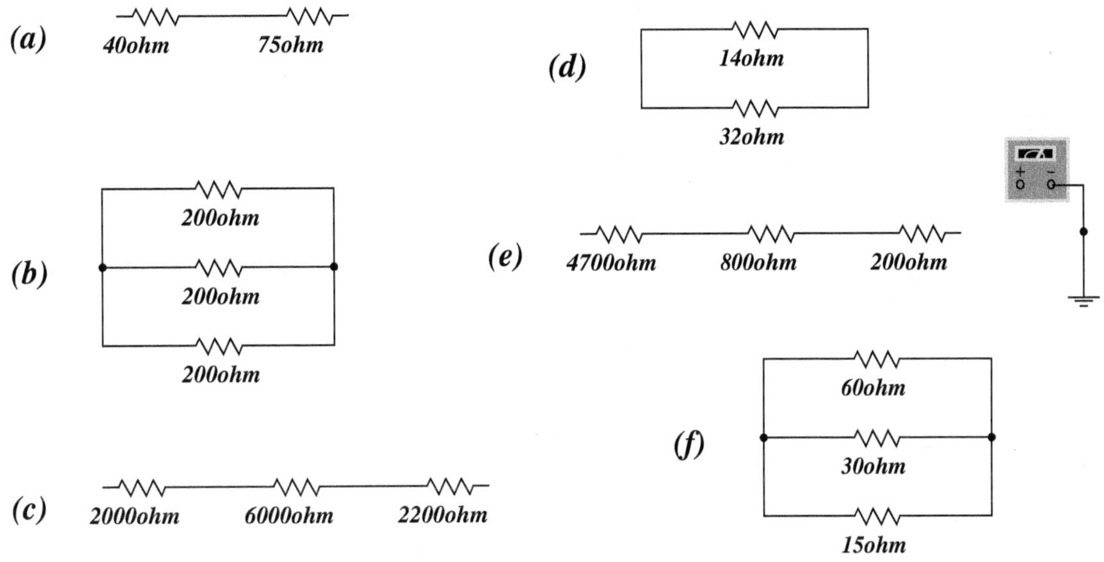

Answers

8-7 Open the Multisim circuit file CH8_07.

(a) **Calculate** the value of I_1 **and** I_2 for the **current-divider** circuit shown.

(b) Turn the simulation on. **Measure** and **record** the value of I_1 **and** I_2.

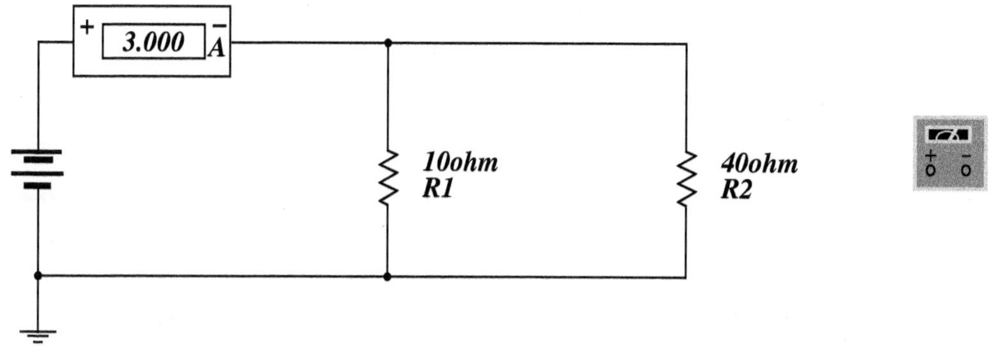

Answers

8-8 Open the Multisim circuit file CH8_08.

(a) **Calculate** the **total resistance** of the **series-parallel** connected resistor circuit shown.

(b) Turn the simulation on. **Measure** and **record** the **total resistance** of the circuit.

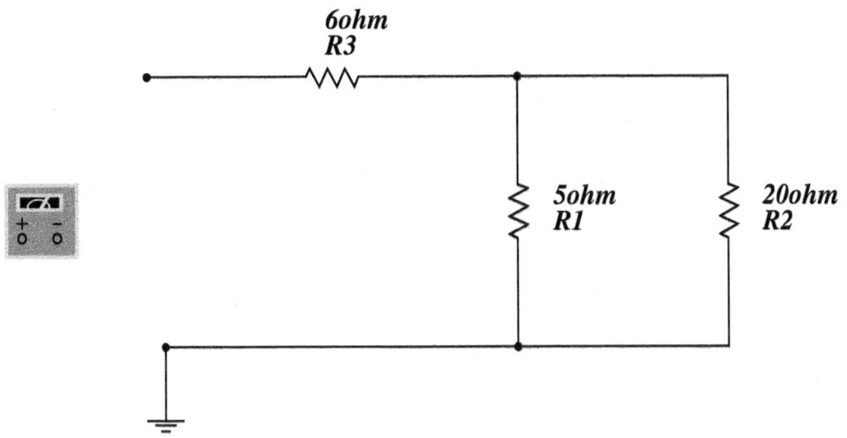

<u>**Answers**</u>

8-9 Open the Multisim circuit file CH8_09.

(a) You have been given three 100 Ω resistors to connect together in three different circuit configurations. What **three circuit configurations** are possible?

(b) **Calculate** the **total resistance** for each circuit.

(c) Turn the simulation on. **Connect** each circuit in turn and record the **total resistance** as **measured** using the multimeter.

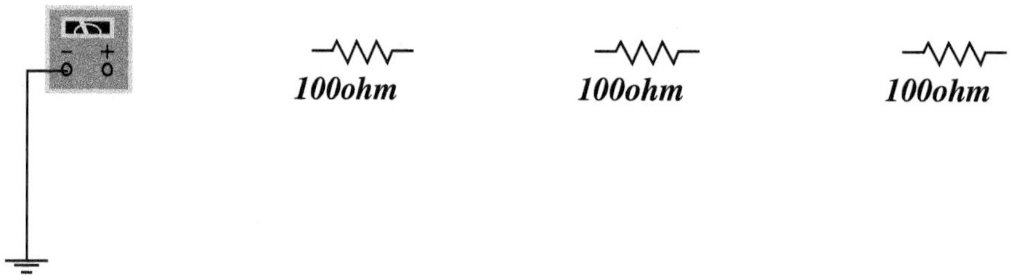

Answers

8-10 Open the Multisim circuit file CH8_10.

(a) **Calculate** the **total resistance** of the resistor network shown.

(b) Turn the simulation on. **Measure** and **record** the total resistance.

(c) Assume resistor is **R_1 burnt open** (infinite resistance). **Calculate** the new value of the total resistance.

(d) **Replace R_1** with the **open** resistor. Turn the simulation on. **Measure** and **record** the total resistance.

(e) Assume resistor **R_3 of the original circuit is short-circuited** (zero resistance). **Calculate** the new value of the total resistance.

(f) **Replace R_3** with the **shorted** resistor. Turn the simulation on. **Measure** and **record** the total resistance.

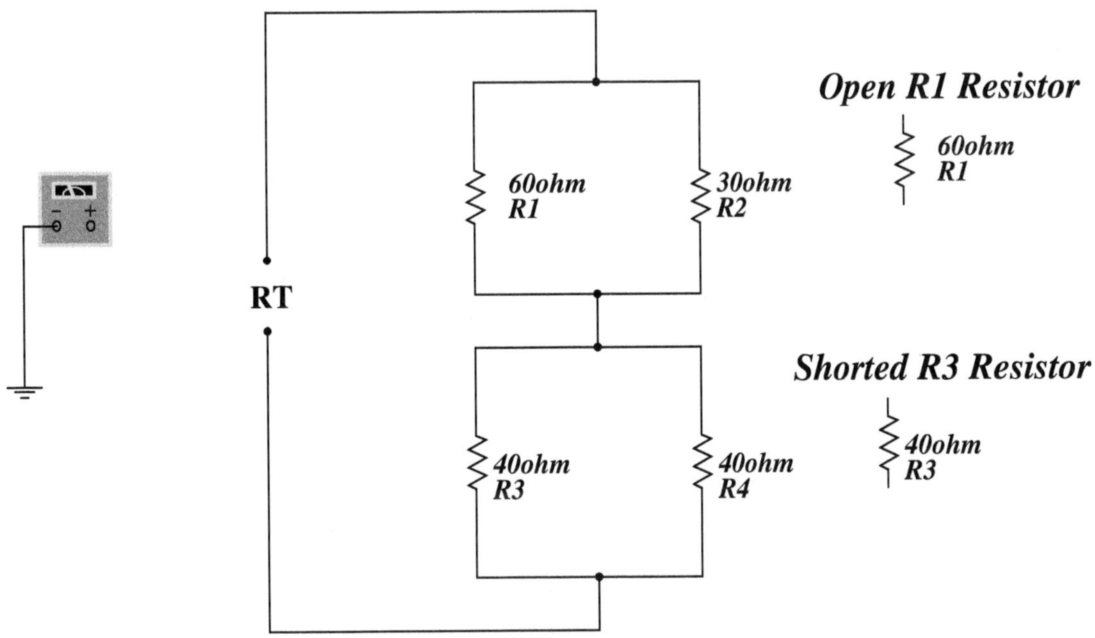

<u>Answers</u>

CHAPTER 9 Electricity and Magnetism

9-1 Open the Multisim circuit file CH9_01.

The circuit shown is that for a **step-down** control **transformer**.

Turn the simulation on. **Measure** and **record**:

(a) the **AC input voltage** applied across the primary
(b) the **AC output voltage** obtained across the secondary coil

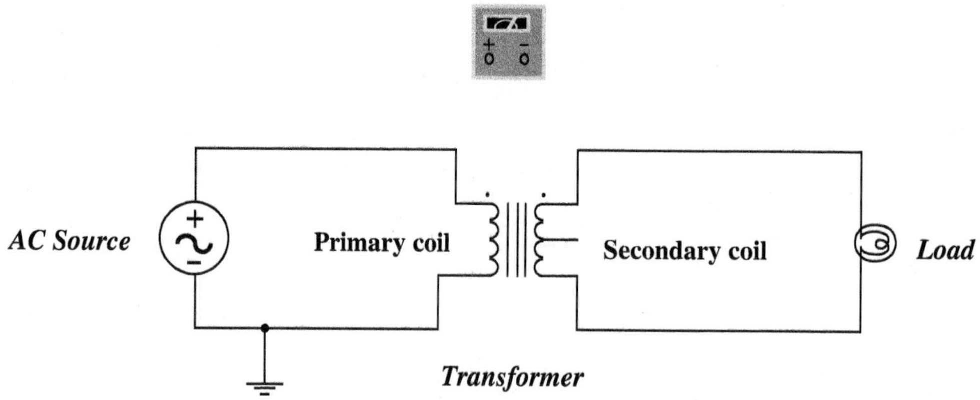

Answers

9-2 Open the Multisim circuit file CH9_02.

The circuit shown is that for an **electromagnetic control relay**.

Turn the simulation on.

(a) Close the switch to establish current flow through the relay coil and **make note** of what happens to the **relay contact.**

(b) **Complete the circuit diagram** of the relay contact connected to **operate the alarm device** any time the **switch is closed**.

(c) **Connect the Multisim circuit** according to the diagram and simulate the normal operation of the circuit.

(d) **Measure** and **record** the AC operating current of the **alarm device.**

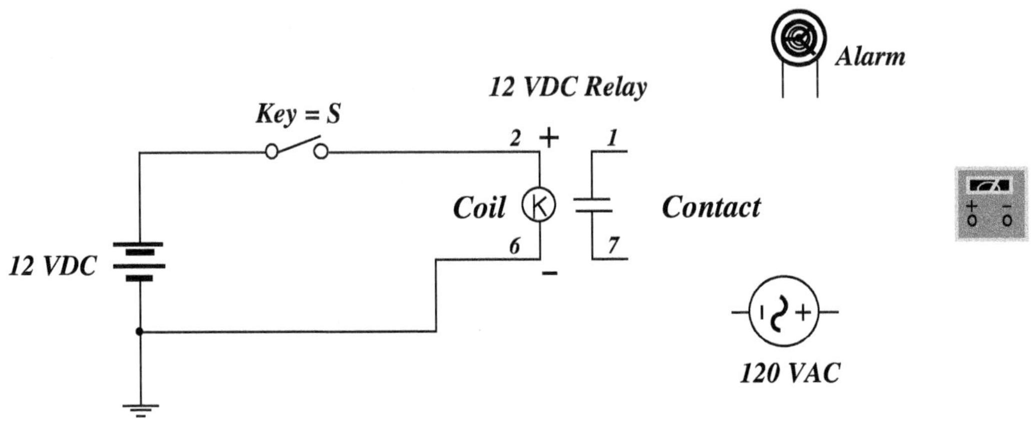

Answers

CHAPTER 10 Electric Power and Energy

10-1 Open the Multisim circuit file CH10_01.

(a) Turn the simulation on. **Record** the amount of **current** drawn by the range element.

(b) **Calculate** the **power** rating of the **range element**.

(c) **Connect the wattmeter** into the Multisim circuit and **record** the amount of **power** being used by the range element.

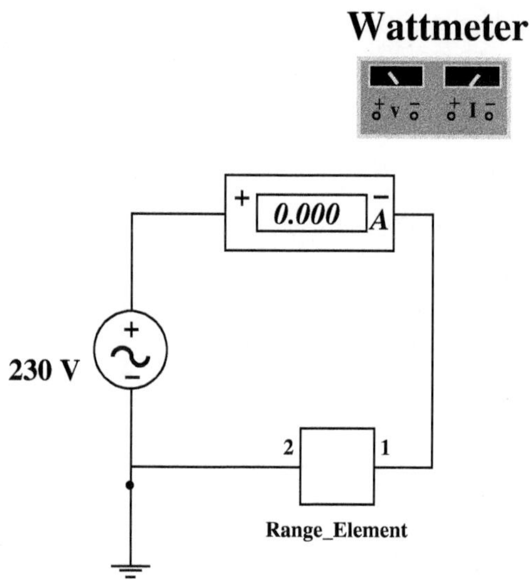

Answers

10-2 Open the Multisim circuit file CH10_02.

(a) Turn the simulation on. **Record** the amount of **power** being used by the **toaster**.

(b) **Calculate** the amount of **current** being drawn by the **toaster**.

(c) **Measure** and **record** the amount of **current** being drawn by the toaster.

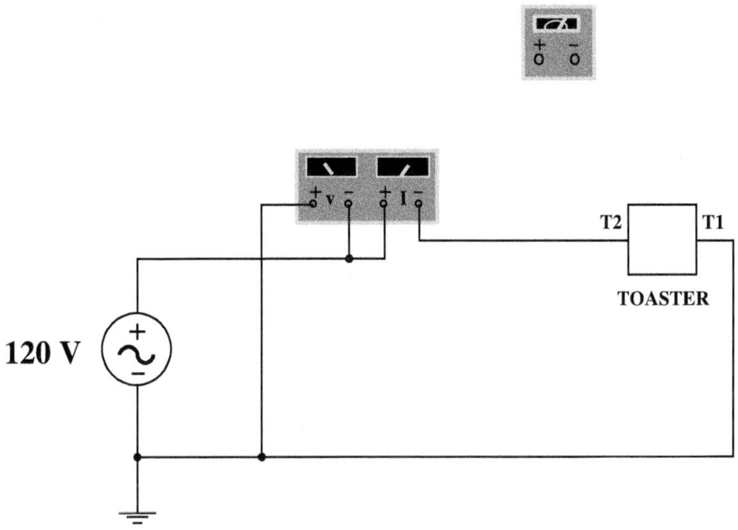

Answers

10-3 Open the Multisim circuit file CH10_03.

(a) Turn the simulation on. **Record** the amount of **current** being drawn by the **resistor**.

(b) **Calculate** the amount of **power** being dissipated in the form of heat from the **resistor.**

(c) **Connect the wattmeter** into the Multisim circuit and **record** the amount of **power** being dissipated by the **resistor**.

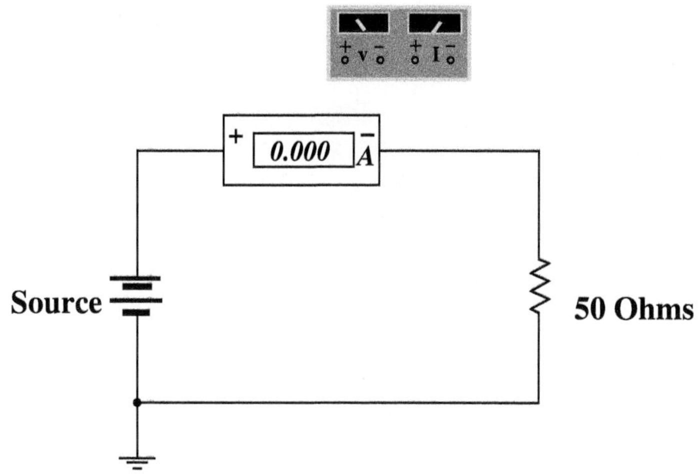

Answers

10-4 Open the Multisim circuit file CH10_04.

(a) **Calculate** the value of the operating **current** for the **40W**, **60W, 150 W**, and **200 W, 120V** light bulbs shown.

(b) Turn the simulation on. **Measure** and **record** the operating **current** of each bulb.

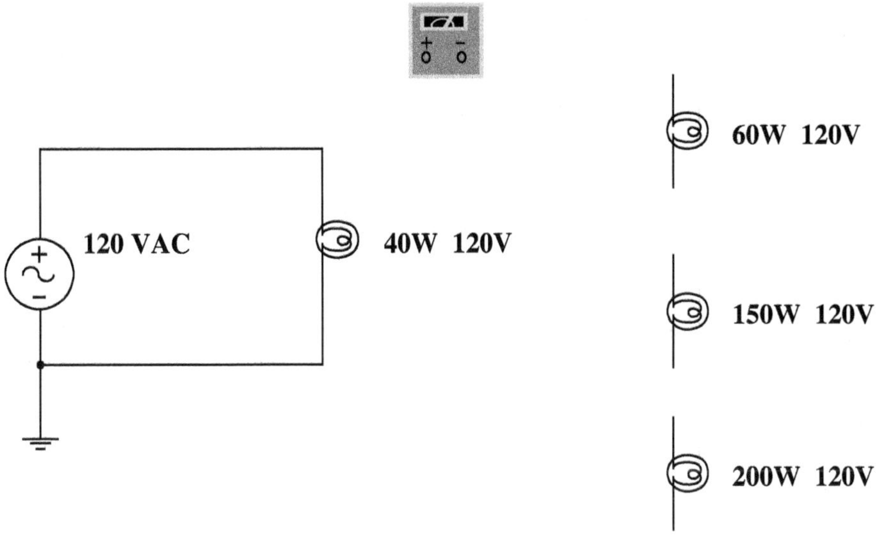

<u>**Answers**</u>

10-5 Open the Multisim circuit file CH10_05.

(a) Turn the simulation on. **Record** the **voltmeter** and **ammeter** readings.

(b) **Calculate** the **total power** of the circuit.

(c) **Connect the wattmeter** and **measure the total power** of the circuit.

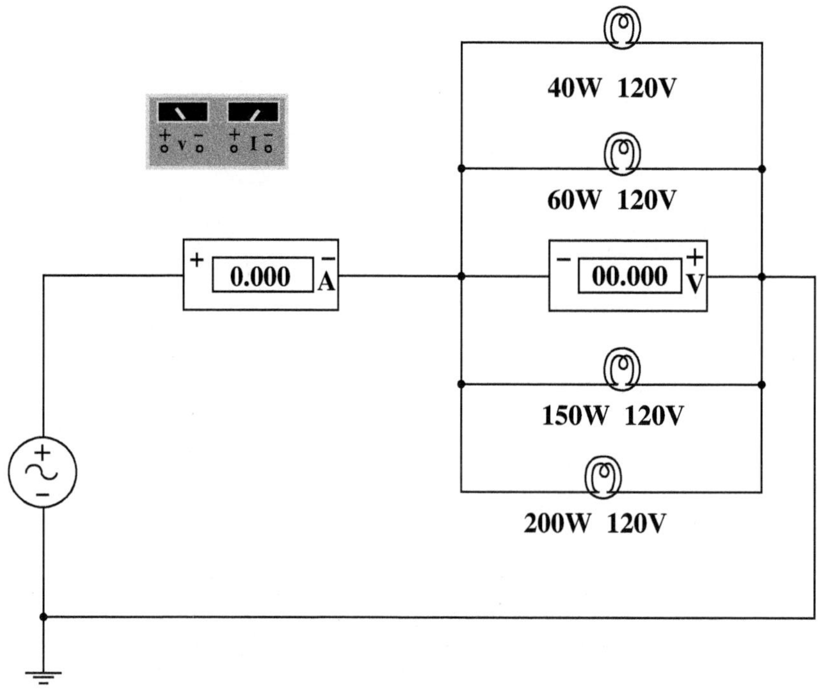

Answers

10-6 Open the Multisim circuit file CH10_06.

(a) Connect the wattmeter to the air conditioner unit and record its operating power.

(b) Calculate the amount of kWh of energy used by the unit if it is operated continuously for 8 hours.

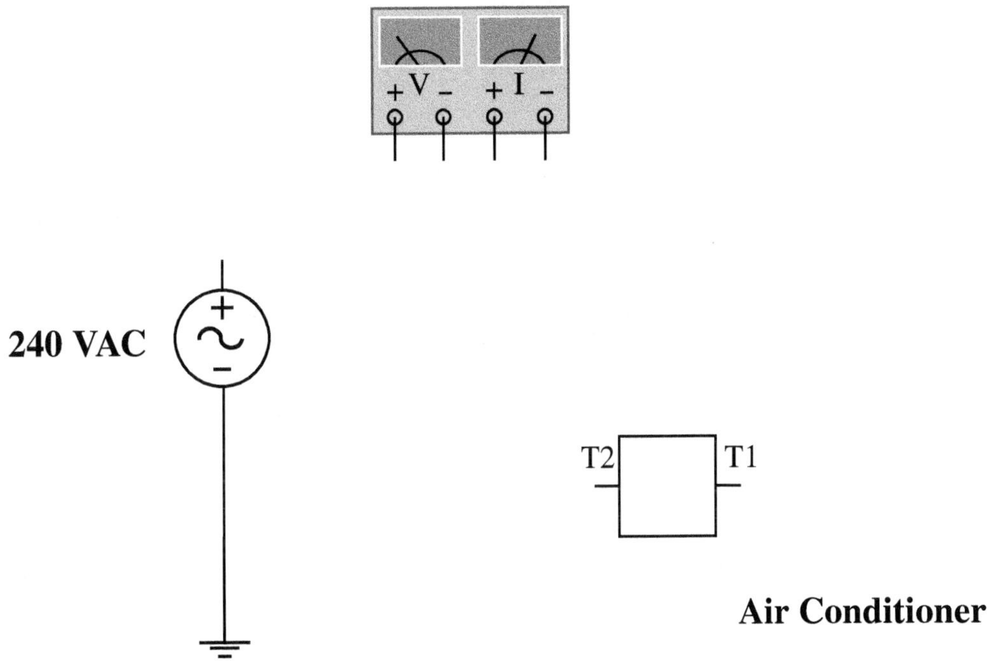

Answers

SECTION **2**

Direct Current (DC) Circuits

Chapter 11	Solving the DC Series Circuit	77
Chapter 12	Solving the DC Parallel Circuit	91
Chapter 13	Solving the DC Series Parallel Circuit	104
Chapter 14	Network Theorems	111

Chapter **11** # Solving the DC Series Circuit

11-1 Open the Multisim circuit file CH11_01.

(a) **Measure** and **record** the **source voltage** and the **voltage across each resistor.**

(b) Based on your readings, **state the voltage characteristics** of a series circuit.

(c) **Measure** and **record** the **source current** and the **current flow through each resistor.**

(d) Based on your current readings, **state the current characteristics** of a series circuit.

(e) **Measure** and **record** the **total circuit resistance** and the **resistance of each resistor**.

(f) Based on your resistance readings, state the **resistance characteristics** of a series circuit.

(g) **Measure** and **record** the **power dissipation of each resistor** and that of the **total circuit**.

(h) Based on your wattmeter readings, state the **power characteristics** of a series circuit.

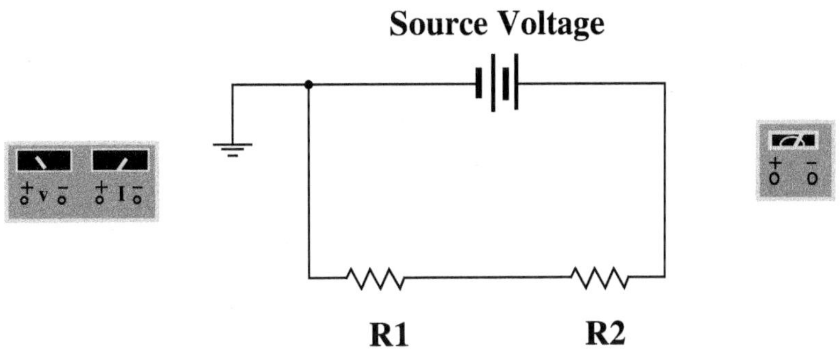

Answers

11-2 Open the Multisim circuit file CH11_02.

(a) **Calculate** the value of the **total resistance.**

(b) **Measure** and **record** the value of the **total resistance**.

(c) **Calculate** the value of the **current flow**.

(d) **Measure** and **record** the value of the **current flow**.

(e) **Calculate** the value of the **voltage across R_1 and R_2**.

(f) **Measure** and **record** the **voltage across R_1 and R_2**.

(g) **State the reason** why the voltage drop across R_1 is equal in value to that across R_2.

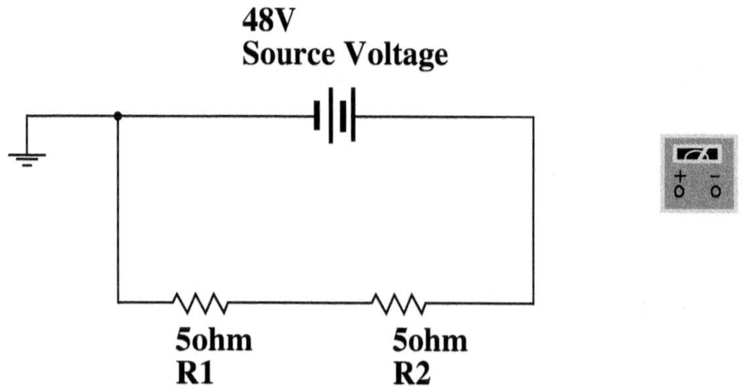

Answers

11-3 Open the Multisim circuit file CH11_03.

(a) **Calculate** the value of the **total resistance.**

(b) **Measure** and **record** the value of the **total resistance**.

(c) **Calculate** the value of the **current flow.**

(d) **Measure** and **record** the value of the **current flow**.

(e) **Calculate** the value of the **voltage across R_1 and R_2**.

(f) **Measure** and **record** the value of the **voltage across R_1 and R_2**.

(g) **State the reason** why the voltage drop across R_2 is greater than that across R_1.

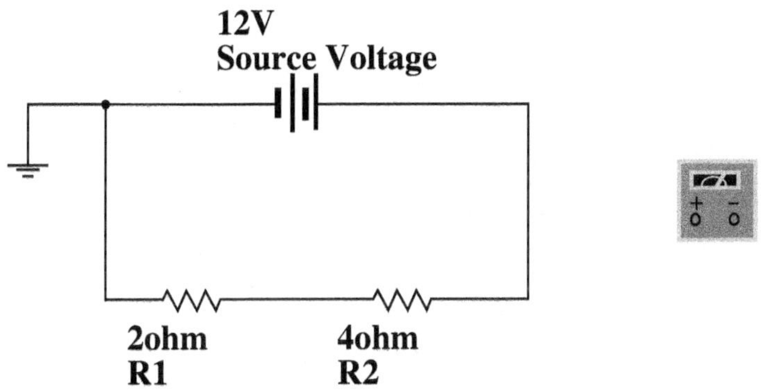

<u>Answers</u>

11-4 Open the Multisim circuit file CH11_04.

(a) **Determine** and **record** (in the table provided) all values of **E, I, R,** and **P** from the measurements given.

(b) Use the multimeter and wattmeter to **measure** and **record** (in the table provided) all values of **E, I, R,** and **P.**

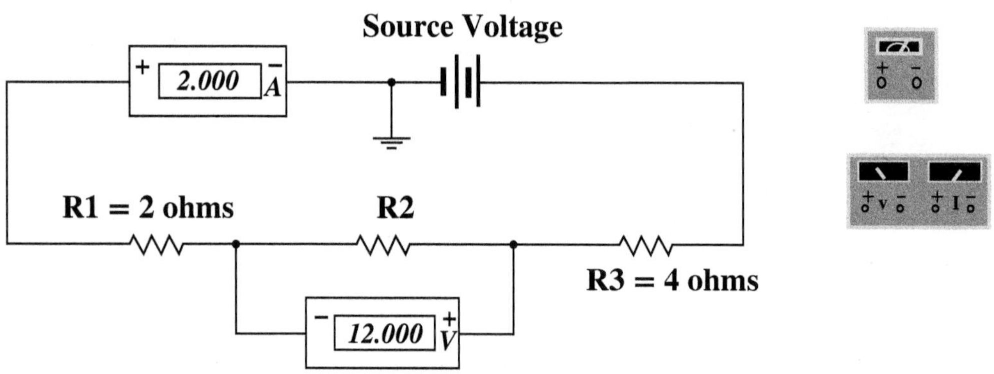

Answers

(a)

	Voltage	Current	Resistance	Power
R_1				
R_2				
R_3				
Total				

(b)

	Voltage	Current	Resistance	Power
R_1				
R_2				
R_3				
Total				

11-5 Open the Multisim circuit file CH11_05.

(a) **Determine** and **record** (in the table provided) all values of **E, I, and R** for the circuit shown.

(b) **Measure** and **record** (in the table provided) all values of **E, I, R, and P.**

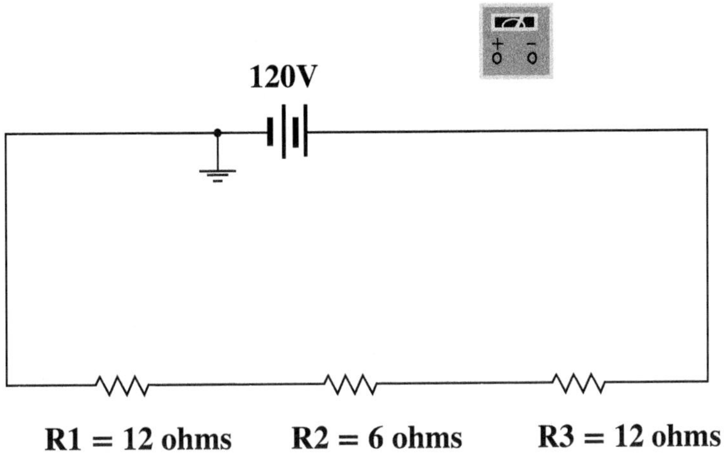

Answers

(a)

	Voltage	Current	Resistance
R_1			
R_2			
R_3			
Total			

(b)

	Voltage	Current	Resistance
R_1			
R_2			
R_3			
Total			

11-6 Open the Multisim circuit file CH11_06.

Measure and **record** the **polarity** (positive or negative) and **value** of the **voltage** from:

(a) **point A** to common ground
(b) **point B** to common ground
(c) **point C** to common ground

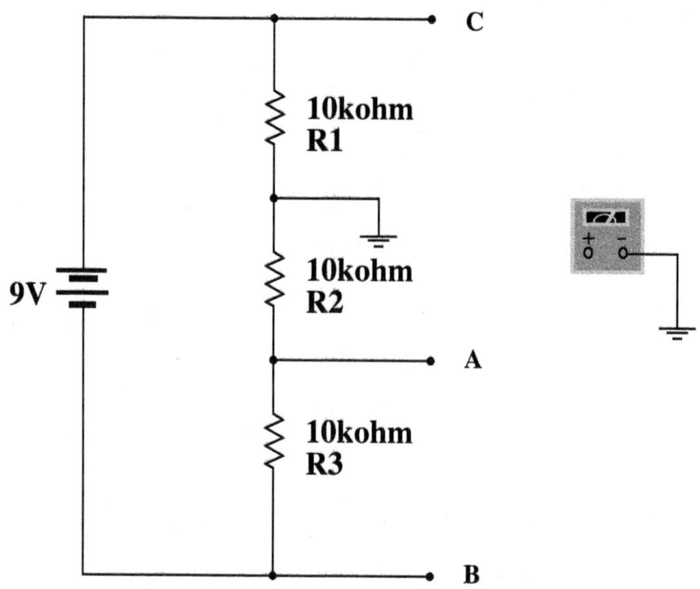

<u>Answers</u>

11-7 Open the Multisim circuit file CH11_07.

(a) **Calculate** the value of the **resistor current** and **voltage** with the batteries connected, as shown, in **series-aiding.**

(b) **Measure** and **record** the **resistor current** and **voltage.**

(c) **Calculate** what the **resistor current** and **voltage** would be with the two batteries connected in **series-opposing.**

(d) **Reconnect** the two batteries in a **series-opposing arrangement. Measure** and **record** the **resistor current** and **voltage.**

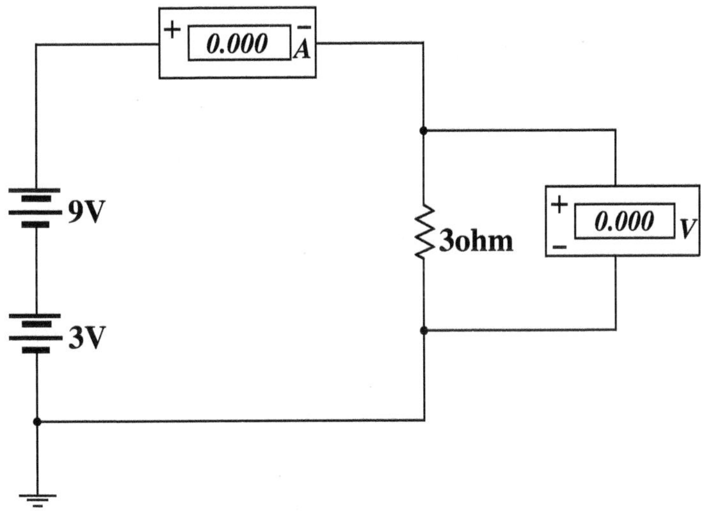

Answers

11-8 Open the Multisim circuit file CH11_08.

(a) Turn the simulation on and **record** the **no-load voltage** of Battery_1.

(b) **Close the switch** and record the **battery voltage and current** with the load connected.

(c) **Disconnect Battery_1** from the circuit and **connect Battery_2** in its place. Turn the simulation on and **record** the **no-load voltage** of Battery_2.

(d) **Close the switch** and record the **battery voltage and current** of Battery_2 with the load connected.

(e) Which battery operates as an ideal **voltage source**? **Why**?

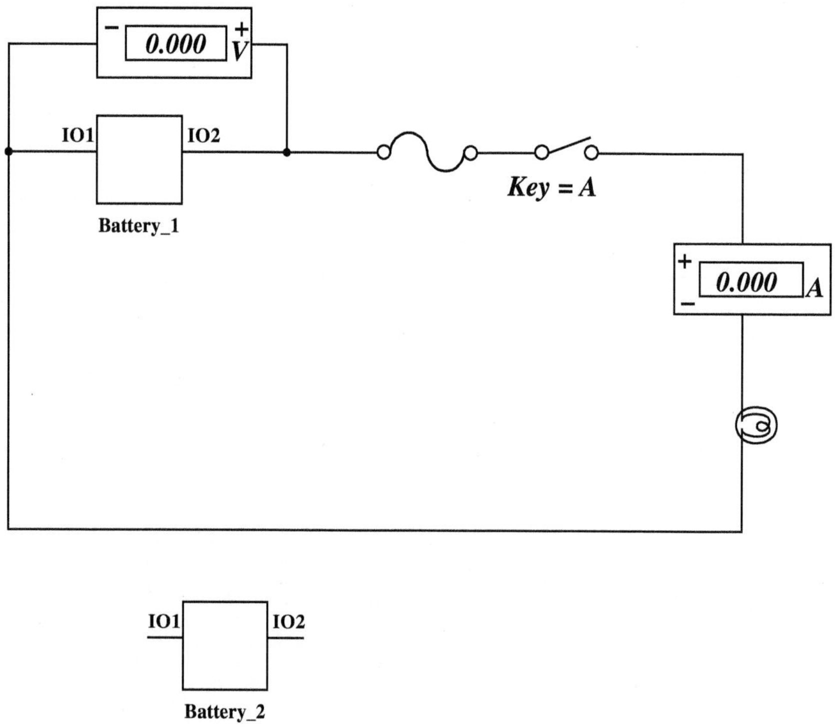

Answers

11-9 Open the Multisim circuit file CH11_09.

(a) Turn the simulation on and **record** the circuit's **load current**.

(b) **Calculate** the **voltage** drop across each of the **line conductors**.

(c) **Measure** and **record** the **voltage drop across** each of the line conductors.

(d) **Calculate** the **voltage** across the **load**.

(e) **Measure** and **record voltage** across the **load**.

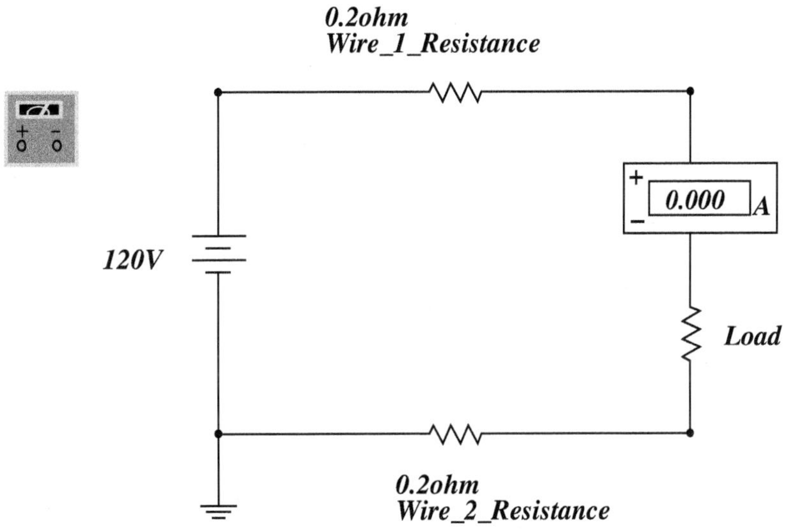

<u>**Answers**</u>

11-10 Open the Multisim circuit file CH11_10.

(a) **Determine** and **record** (in the table provided) all values of **E, I, and R** for the circuit shown. Use the **multimeter** to **measure** all values of E, I, R, in order to **verify** your calculations.
(b) **Repeat for the original circuit** using the defective **open resistor R_2** in place of the original one.
(c) **Repeat for the original circuit** using the **defective shorted resistor R_1** in place of the original one.

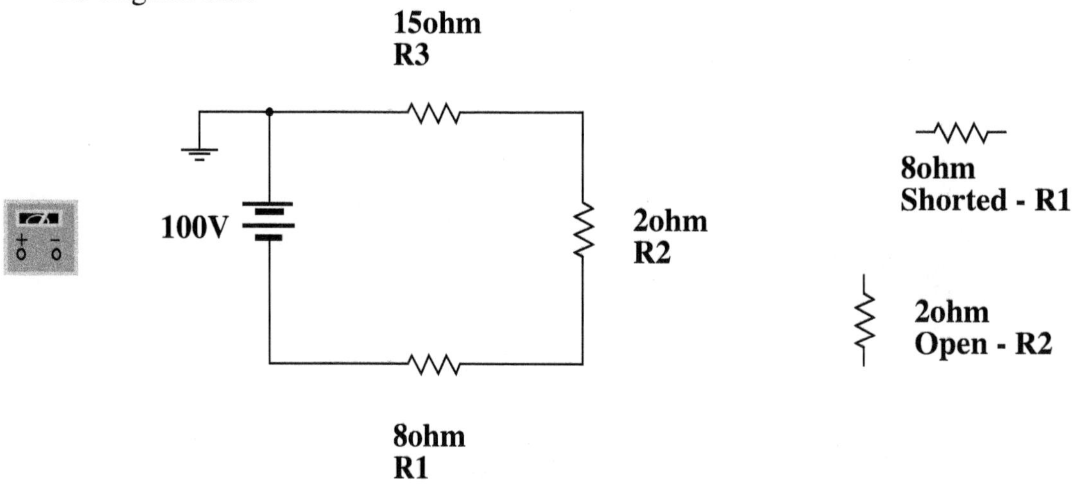

Answers

(a)

	Voltage	Current	Resistance
R_1			
R_2			
R_3			
Total			

(b)

	Voltage	Current	Resistance
R_1			
R_2			
R_3			
Total			

(c)

	Voltage	Current	Resistance
R_1			
R_2			
R_3			
Total			

11-11 Open the Multisim circuit file CH11_11.

Turn the simulation on. Simulate an **open fault in R_1** by **opening** the key **switch. Record** the **change** in:

(a) **current** flow

(b) voltage $\mathbf{E_1}$

(c) voltage $\mathbf{E_2}$

(d) voltage $\mathbf{E_3}$

(e) total **power**

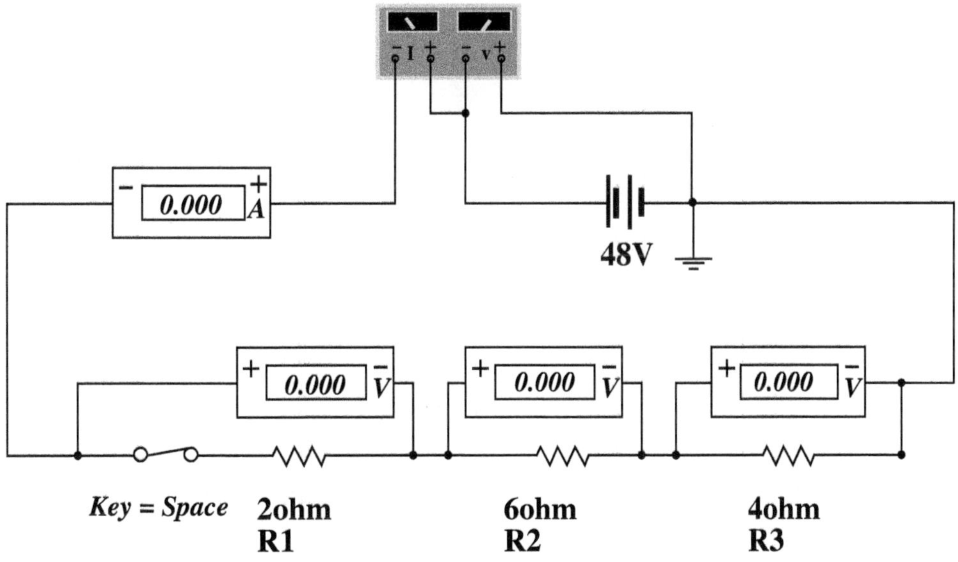

<u>Answers</u>

11-12 Open the Multisim circuit file CH11_12.

Turn the simulation on. Simulate a **shorted fault in R_2** by **closing** the key **switch**. **Record** the **change** in:

(a) **current** flow

(b) voltage E_1

(c) voltage E_2

(d) voltage E_3

(e) total **power**

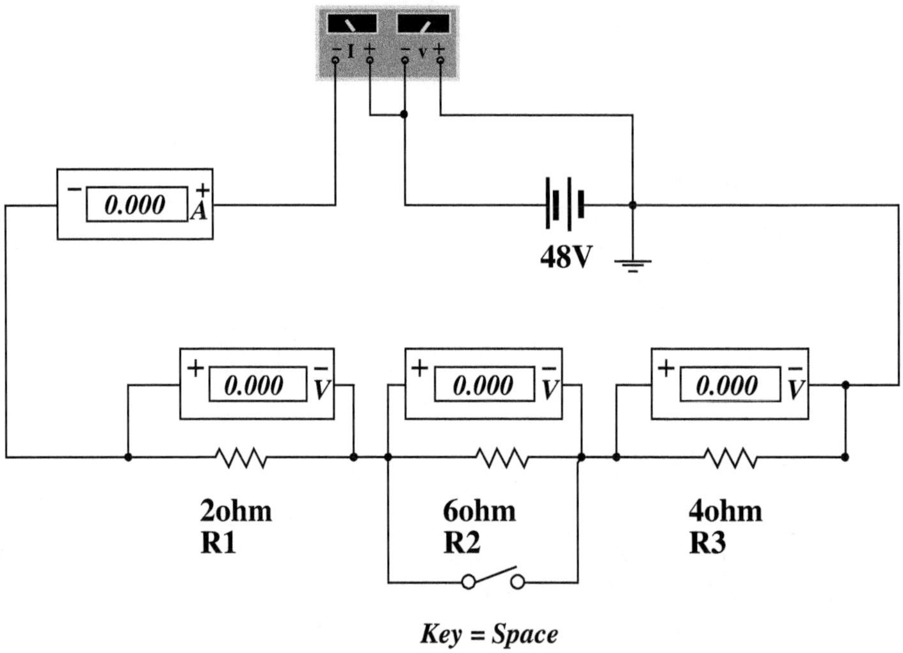

Answers

11-13 Open the Multisim circuit file CH11_13.

(a) Turn the simulation on, and using **only the voltmeter**, determine which circuit is operating **normal** and which circuits are **faulted**.

(b) **Identify** the faulted components and type of fault (**open** or **short**).

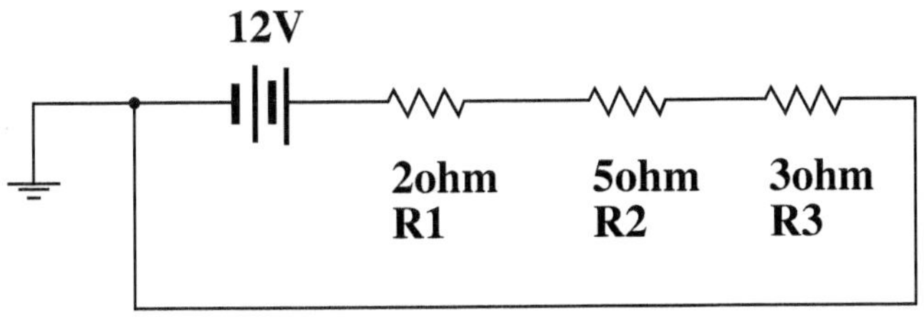

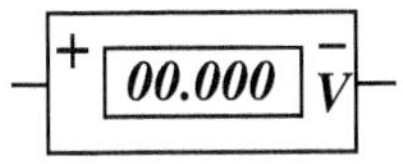

Circuit #1

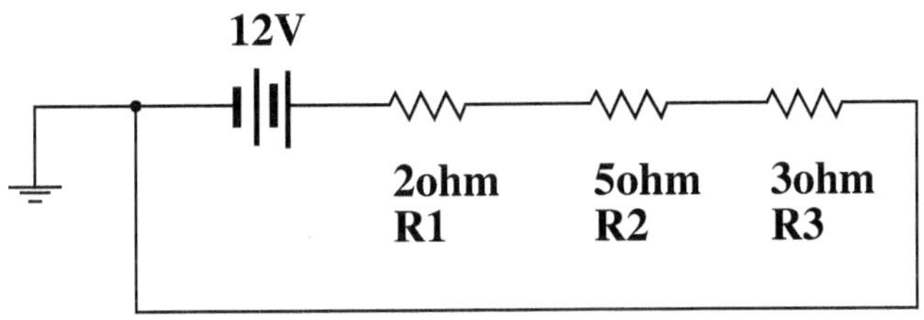

Circuit #2

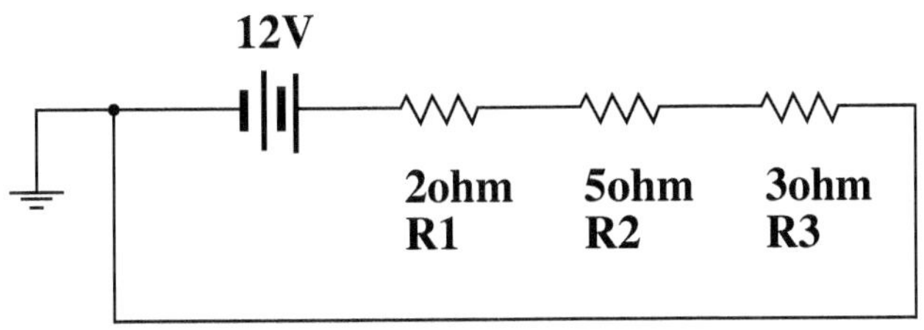

Circuit #3

<u>**Answers**</u>

12-2 Open the Multisim circuit file CH12_02.

(a) **Calculate** the value of the **total resistance.**

(b) **Measure** and **record** the value of the **total resistance.**

(c) **Calculate** the value of the **total current flow** and the **current flow through each of the resistors.**

(d) **Measure** and **record** value of the **total current flow** and the **current flow through each of the resistors.**

(e) What is the **value of the voltage across R_1 and R_2**? Measure these voltages to verify your answer.

(f) **State the reason** why the current flow through R_1 is **equal** in value to that through R_2?

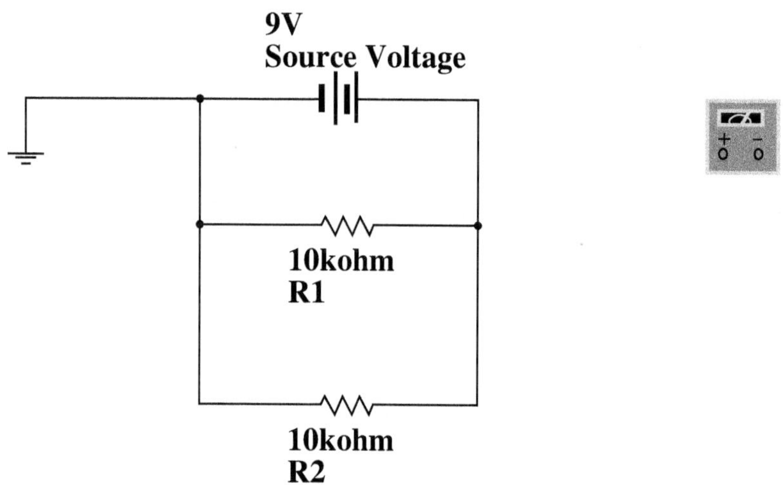

<u>**Answers**</u>

12-3 Open the Multisim circuit file CH12_03.

(a) **Calculate** the value of the **total resistance.**

(b) **Measure** and **record** the value of the **total resistance.**

(c) **Calculate** the value of the **total current flow** and the **current flow through each of the resistors.**

(d) **Measure** and **record** the **total current flow** and the **current flow through each of the resistors.**

(e) What is the **value of the voltage across R_1 and R_2**? Measure these voltages to verify your answer.

(f) **State the reason why** the current flow through R_1 greater than that through R_2.

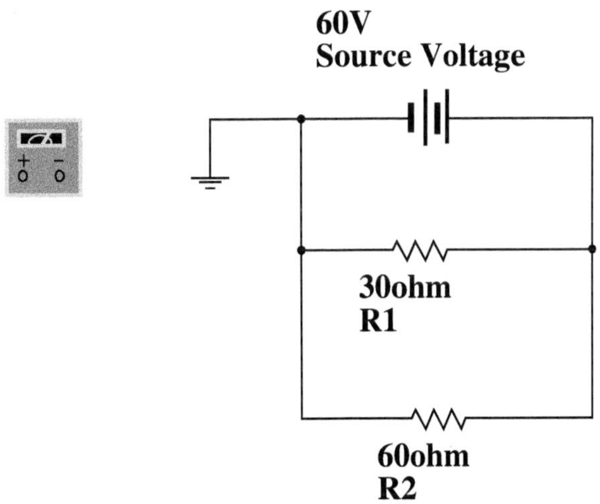

<u>**Answers**</u>

12-4 Open the Multisim circuit file CH12_04.

(a) **Determine** and **record** (in the table provided) all values of **E, I, R, and P** for the circuit shown.

(b) Use the multimeter and wattmeter to **measure** and **record** (in the table provided) all values of **E, I, R, and P.**

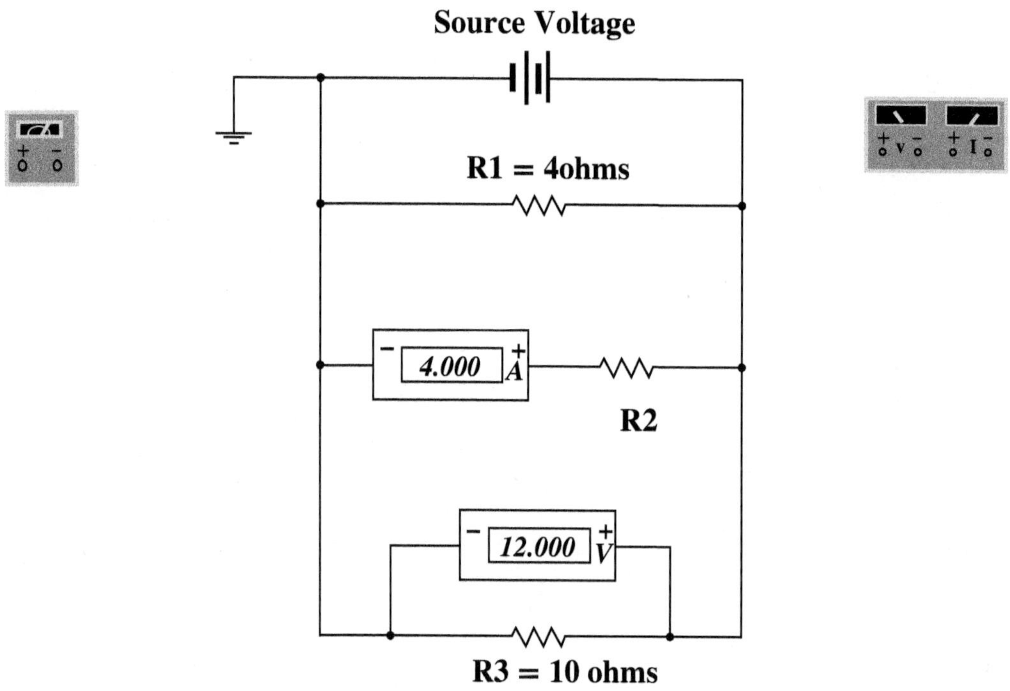

<u>**Answers**</u>

(a)

	Voltage	Current	Resistance	Power
R_1				
R_2				
R_3				
Total				

(b)

	Voltage	Current	Resistance	Power
R_1				
R_2				
R_3				
Total				

12-5 Open the Multisim circuit file CH12_05.

(a) **Determine** and **record** (in the table provided) all values of **E, I, R, and P** for the circuit shown.

(b) Use the multimeter and wattmeter to **measure** and **record** (in the table provided) all values of **E, I, R, and P.**

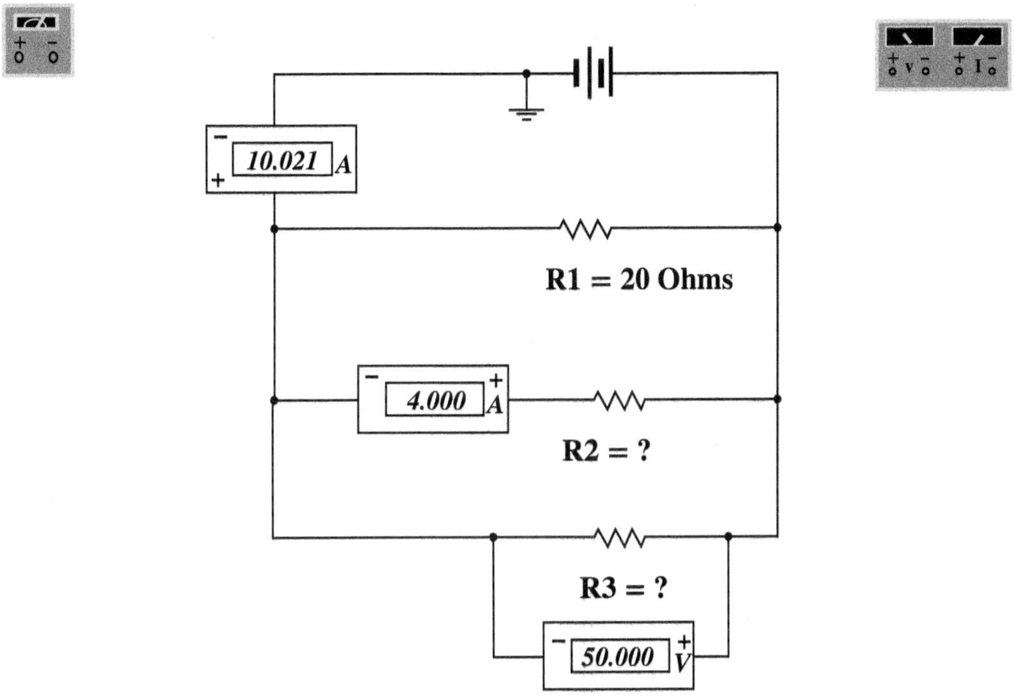

Answers

(a)

	Voltage	Current	Resistance	Power
R_1				
R_2				
R_3				
Total				

(b)

	Voltage	Current	Resistance	Power
R_1				
R_2				
R_3				
Total				

12-6 Open the Multisim circuit file CH12_06.

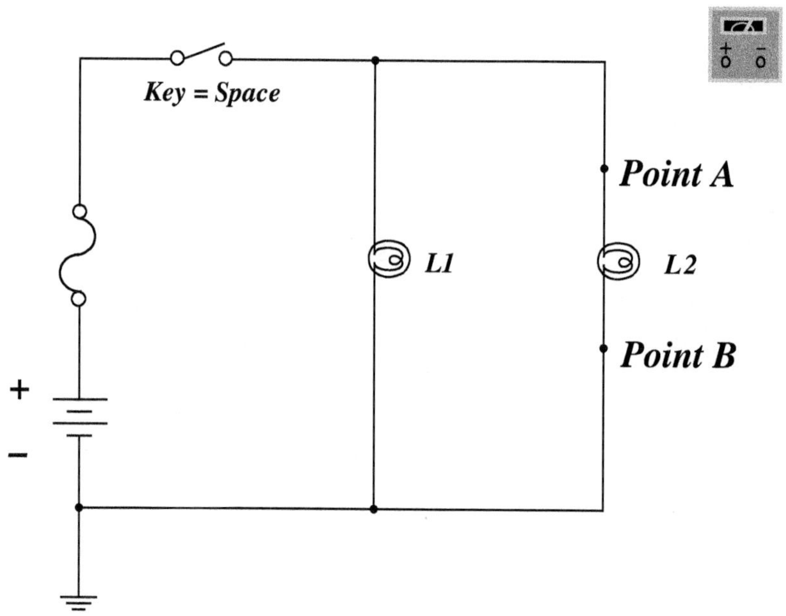

Answer

12-7 Open the Multisim circuit file CH12_07.

(a) **Determine** and **record** (in the table provided) all values of **E, I, R, and P** for the circuit shown. Use the multimeter to **verify your answers.**

(b) **Determine** and **record** (in the table provided) all values of **E, I, R, and P** using the **faulted open resistor R_1**. Use the multimeter to **verify your answers.**

(c) Operate the simulation with the **faulted shorted resistor R_1** connected. **What happens? Why?**

(d) **Record** the total **resistance** of the circuit with the **faulted shorted** resistor R_1 connected.

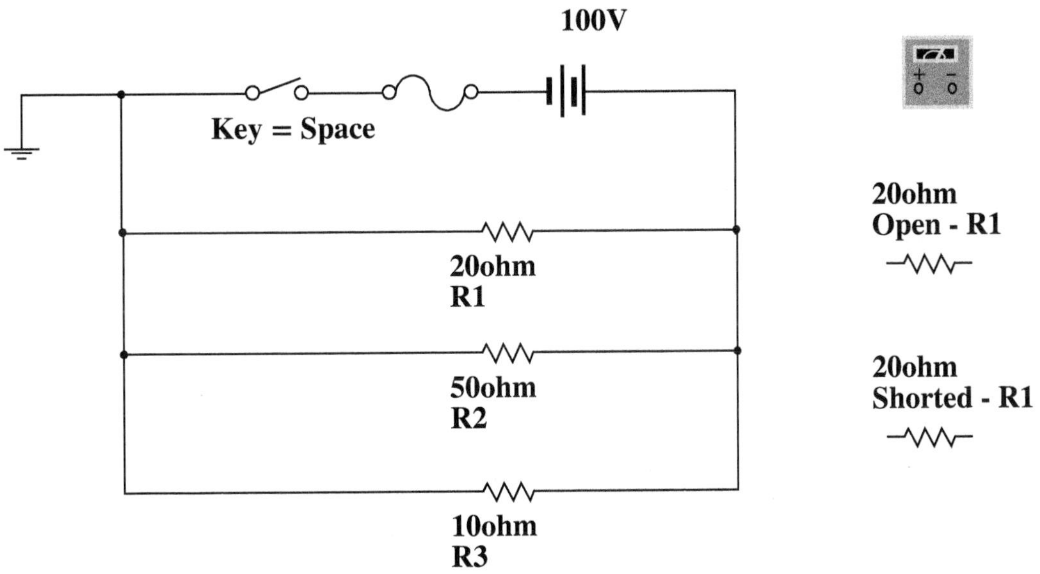

<u>**Answers**</u>

(a)

	Voltage	Current	Resistance
R_1			
R_2			
R_3			
Total			

(b)

	Voltage	Current	Resistance
R_1			
R_2			
R_3			
Total			

12-8 Open the Multisim circuit file CH12_08.

Turn the simulation on. Simulate an **open fault in R_1** by **opening** the key **switch. Record** the **change** in:

(a) I_T
(b) I_1
(c) I_2
(d) E_T
(e) E_1
(f) E_2
(g) P_T

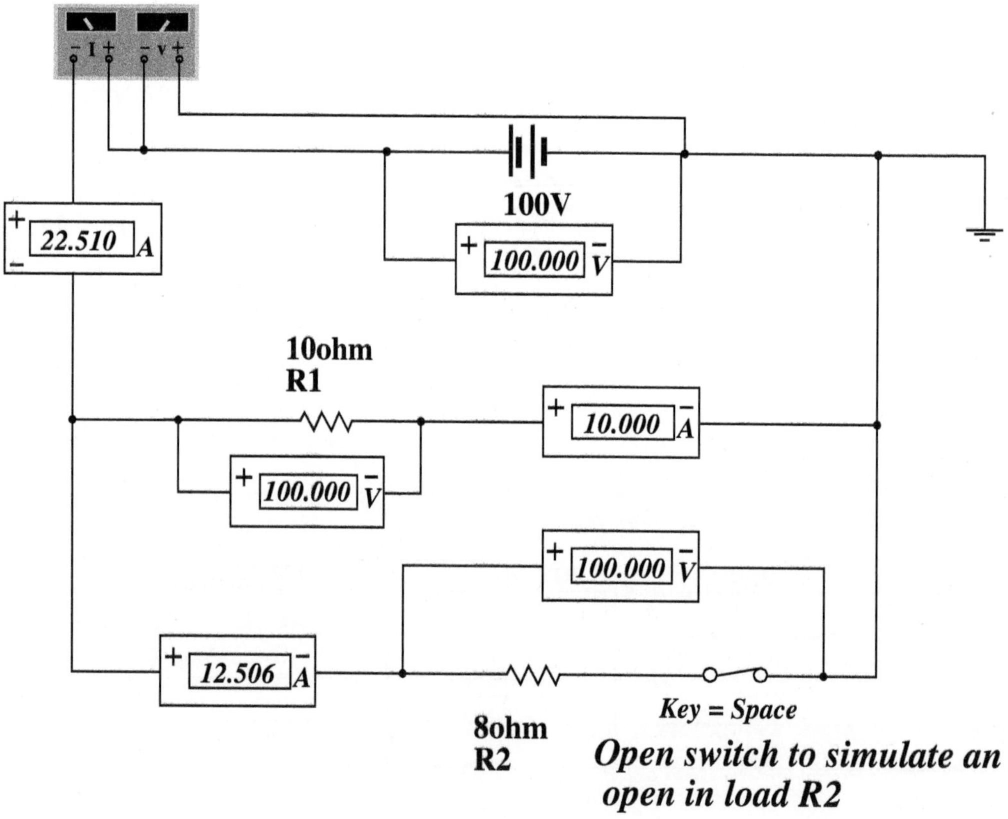

<u>Answers</u>

12-9 Open the Multisim circuit file CH12_09.

(a) Turn the simulation on. Using the **multimeter,** determine which circuit is **operating normally** and which circuits are **at fault.**

(b) **Identify** the faulted components and type of fault (**open or short**).

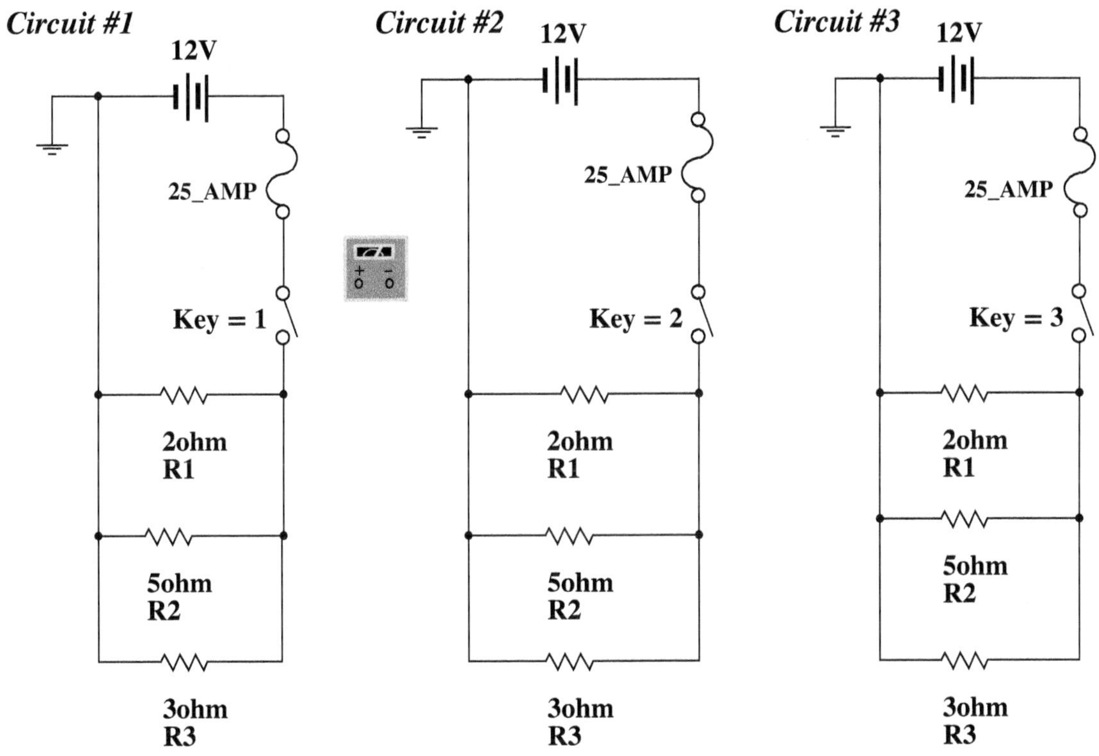

Answers

Chapter 13 Solving the DC Parallel Series–Parallel Circuit

13-1 Open the Multisim circuit file CH13_01.

(a) **Determine** and **record** (in the table provided) all values of **E, I, and R** from the measurements given.

(b) Use the multimeter to **measure** and **record** (in the table provided) all values **of E, I, and R.**

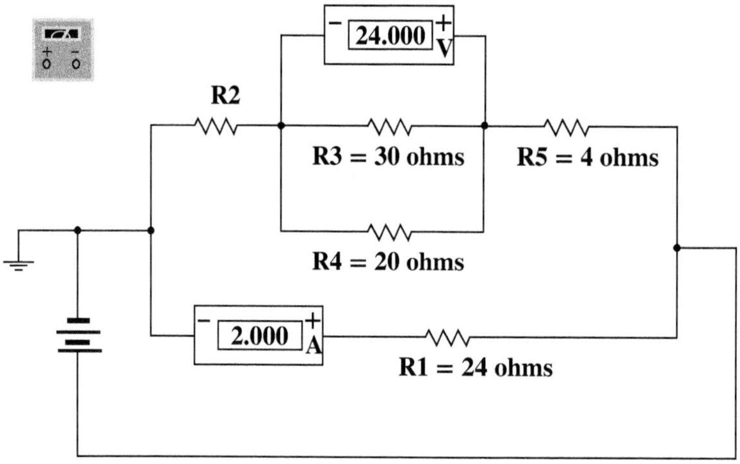

Answers

(a)

	Voltage	Current	Resistance
R_1			
R_2			
R_3			
R_4			
R_5			
Total			

(b)

	Voltage	Current	Resistance
R_1			
R_2			
R_3			
R_4			
R_5			
Total			

13-2 Open the Multisim circuit file CH13_02.

Kirchhoff's current law states that the current entering any point in a circuit is equal to the current leaving that same point.

Kirchhoff's voltage law states that the sum of the voltage drops around a single closed loop of a circuit must be equal to that of the source voltage.

(a) Use **Kirchhoff's current and voltage laws** to **determine** and **record** I_1**,** E_2**,** I_4**, and** E_4 for the series-parallel circuit.

(b) **Measure** and **record** the value of I_1**,** E_2**,** I_4**, and** E_4 to verify your answers.

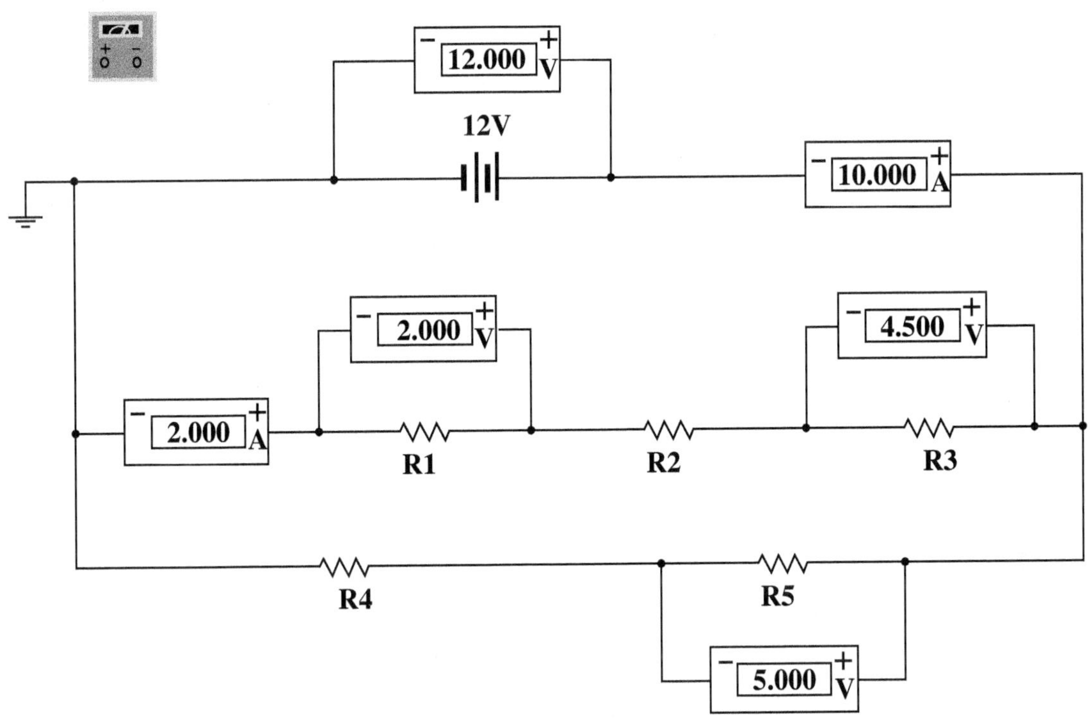

<u>**Answers**</u>

13-3 Open the Multisim circuit file CH13_03.

(a) **Determine** and **record** (in the table provided) all values of **E, I, R, and P** for the circuit shown.

(b) Use the multimeter and wattmeter to **measure** and **record** (in the table provided) all values of E, I, R, and P.

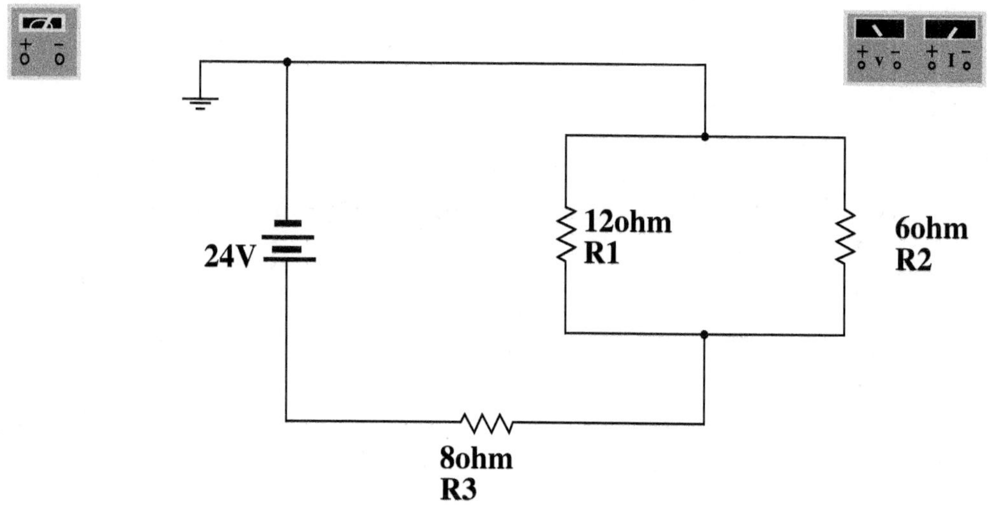

Answers

(a)

	Voltage	Current	Resistance	Power
R_1				
R_2				
R_3				
Total				

(b)

	Voltage	Current	Resistance	Power
R_1				
R_2				
R_3				
Total				

13-4 Open the Multisim circuit file CH13_04.

(a) **Determine** and **record** (in the table provided) all values of **E, I,** and **R** for the circuit. Use the multimeter to **verify your answers.**

(b) **Repeat** for the original circuit using the **faulted open resistor R_3**.

(c) **Repeat** for the original circuit using the **faulted shorted resistor R_1.**

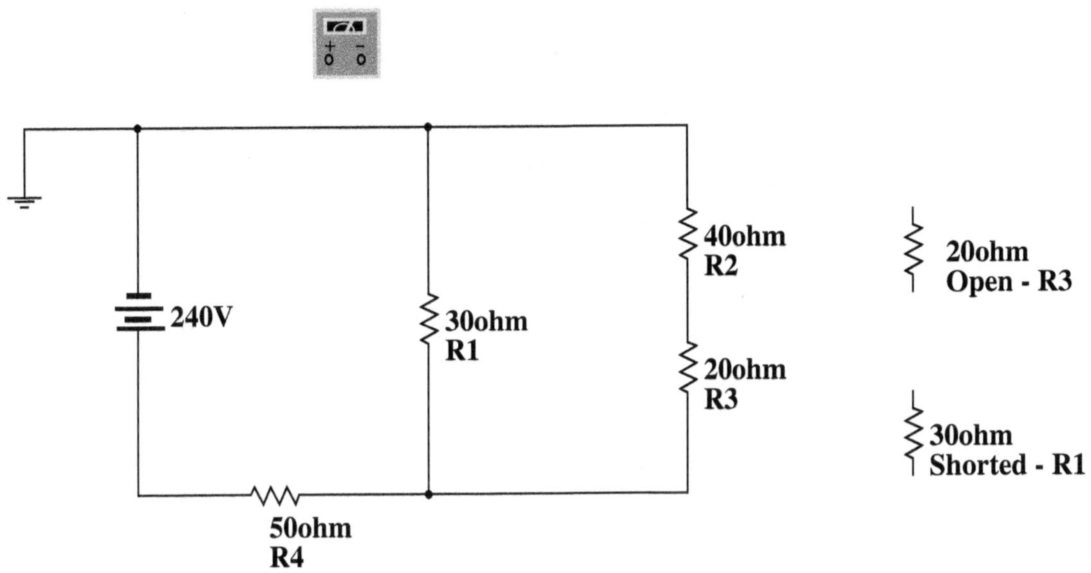

Answers

(a) Normal circuit.

	Voltage	Current	Resistance
R_1			
R_2			
R_3			
R_4			
Total			

(b) Faulted open resistor R_3 circuit.

	Voltage	Current	Resistance
R_1			
R_2			
R_3			
R_4			
Total			

(c) Faulted shorted resistor R_1 circuit.

	Voltage	**Current**	**Resistance**
R_1			
R_2			
R_3			
R_4			
Total			

13-5 Open the Multisim circuit file CH13_05.

Measure and **record** the value of the **voltage** drop across each of the resistor loads. Based on your voltage readings, **which resistor is at fault**? Give a **reason** for your answer and classify the fault condition as an **open** or **short.**

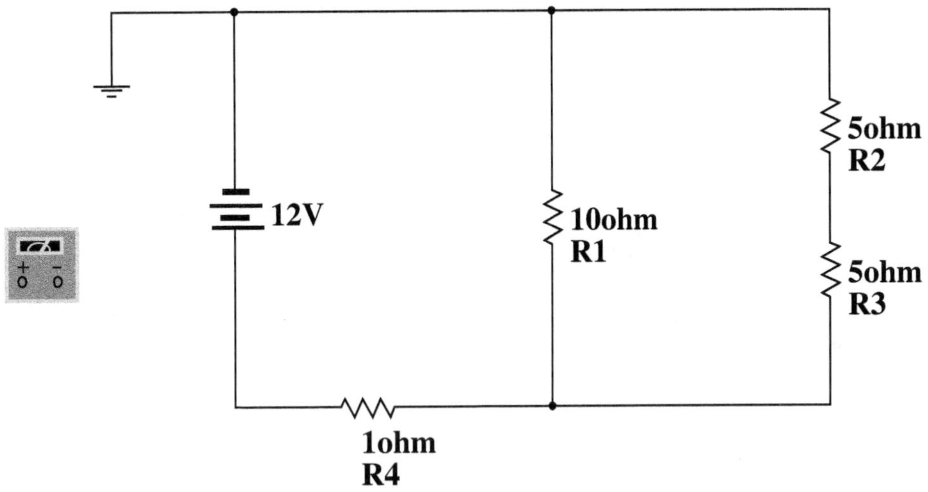

<u>**Answers**</u>

13-6 Open the Multisim circuit file CH13_06.

(a) **Measure** and **record** the value of the **voltage** drop across each of the resistor loads.

(b) Based on your voltage readings, **which resistor is at fault**? Give a **reason** for your answer and classify the fault condition as an **open** or **short.**

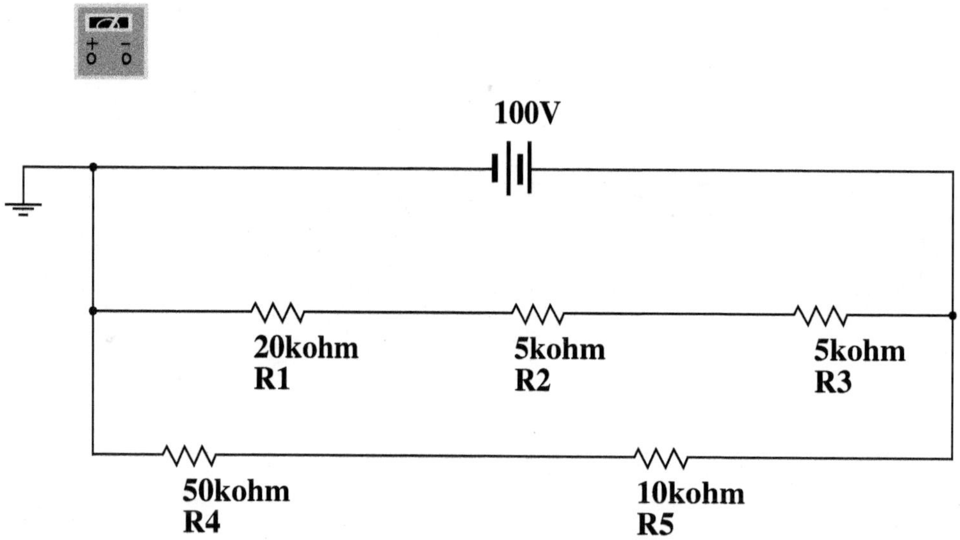

Answers

Chapter 14 Network Theorems

14-1 Open the Multisim circuit file CH14_01.

(a) **Calculate** and **record** the output voltage of the **real voltage source** shown with the **R_L 5Ω load** connected. Measure the amount of output voltage to verify your calculation.

(b) **Repeat** using the **R_L 10Ω load** resistor.

(c) **Repeat** using the **R_L 2Ω load** resistor.

(d) What happens to output voltage (**greater or less output**) when the **load** on the circuit is **increased** (less resistance) and **decreased** (more resistance)?

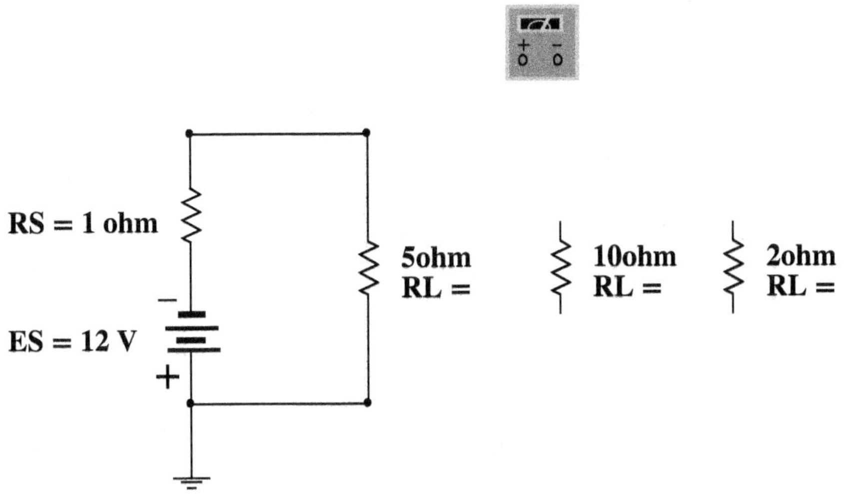

<u>Answers</u>

14-2 Open the Multisim circuit file CH14_02.

(a) **Calculate** and **record** the amount of current flow through R_L produced by the **current source**, with the **R_L 5Ω load** connected. Measure the amount current of flow through R_L to verify your calculation.

(b) Repeat using the **R_L 10Ω load** resistor.

(c) Repeat using the **R_L 2Ω load** resistor.

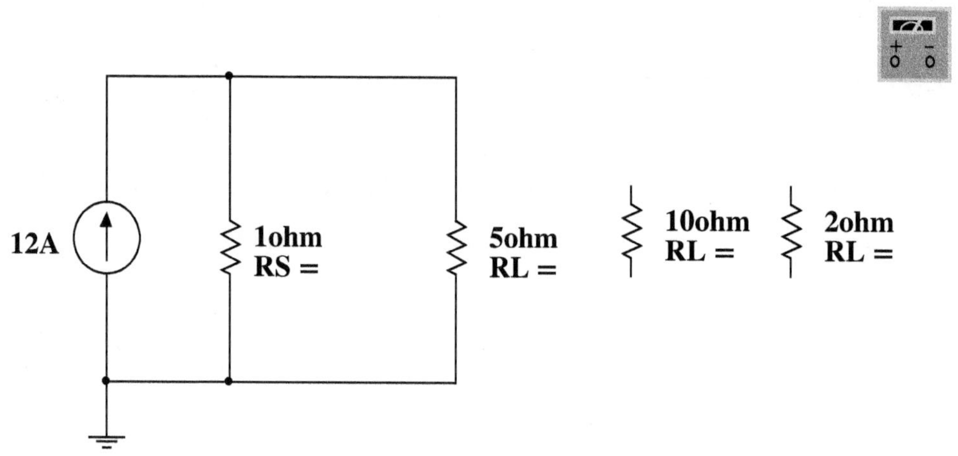

<u>Answers</u>

14-3 Open the Multisim circuit file CH14_03.

(a) **Calculate** and **record** the amount of **current** flow through and **voltage** drop across **R_1 and R_2** using the **superposition theorem.**

(b) **Verify** your answers by **measuring** and **recording** the current and voltage of R_1 and R_2.

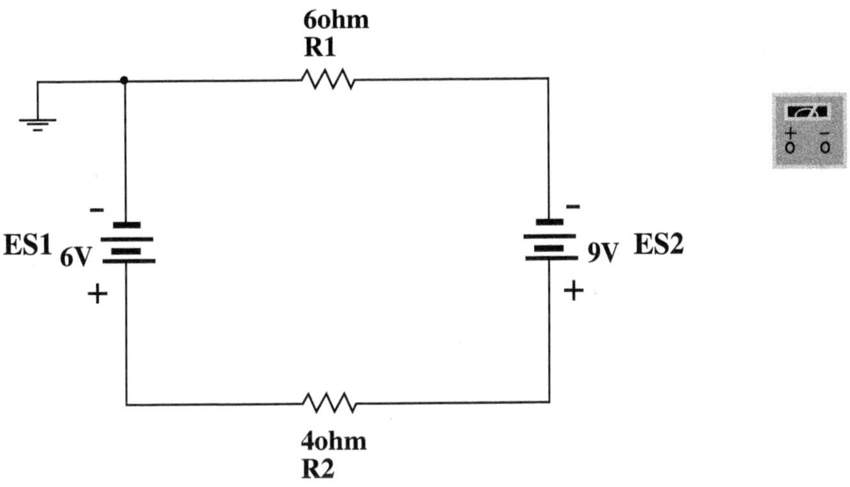

<u>Answers</u>

14-4 Open the Multisim circuit file CH14_04.

(a) **Calculate** and **record** the amount of **current** flow through $\mathbf{R_3}$ using the **superposition theorem.**

(b) **Verify** your answer by **measuring** and **recording** the amount of **current** flow through $\mathbf{R_3}$.

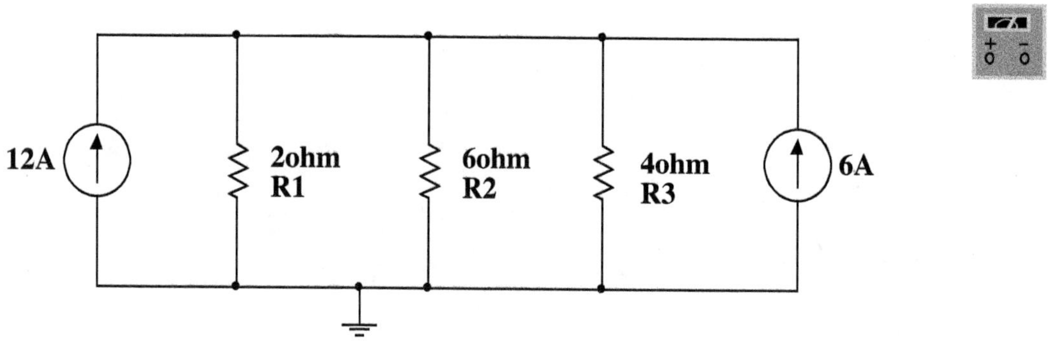

<u>Answers</u>

14-5 Open the Multisim circuit file CH14_05.

(a) **Calculate** and **record** current flow $\mathbf{I_1}$, $\mathbf{I_2}$, **and** $\mathbf{I_3}$ using the **superposition theorem.**

(b) **Verify** your answers by **measuring** and **recording** the amount of current that flows through $\mathbf{I_1}$, $\mathbf{I_2}$, **and** $\mathbf{I_3}$.

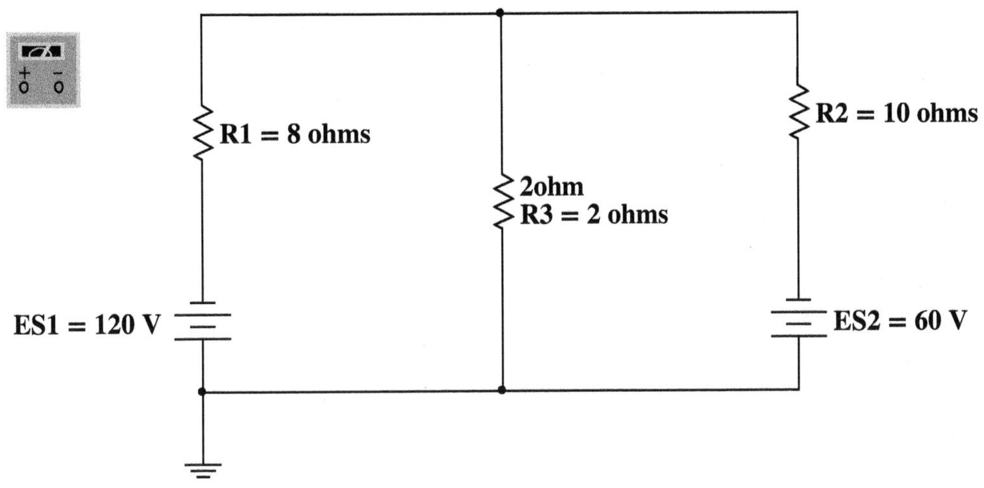

<u>Answers</u>

14-6 Open the Multisim circuit file CH14_06.

(a) **Calculate** and **record** the amount of **current** flow through $\mathbf{R_2}$ using the **superposition theorem.**

(b) **Verify** your answer by **measuring** and **recording** the current flow through $\mathbf{R_2}$.

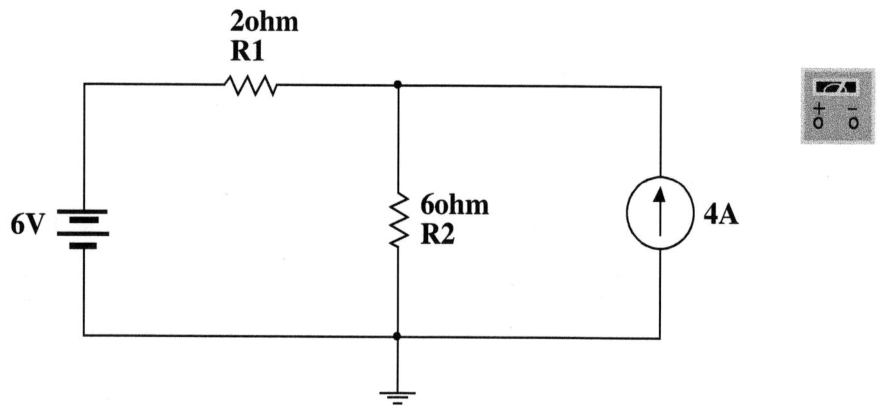

Answers

14-7 Open the Multisim circuit file CH14_07.

(a) **Calculate** and **record** the amount of **current** flow through the **common wire** using the **superposition theorem. Verify** your answer by **measuring** the current flow through the **common wire.**

(b) **Repeat** for an $\mathbf{R_2}$ **value of 20Ω.**

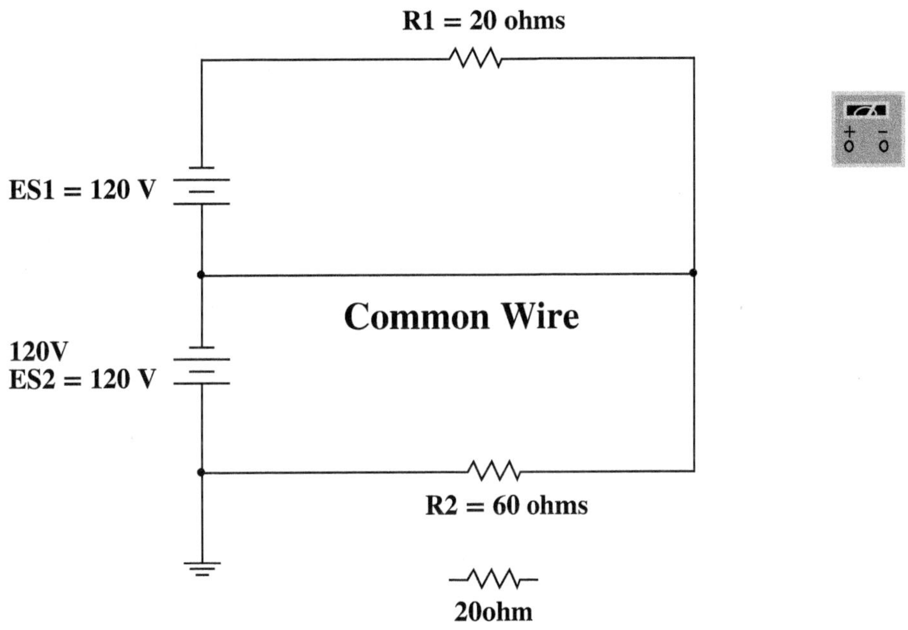

Answers

14-8 Open the Multisim circuit file CH14_08.

(a) **Calculate** and **record** the **current** flow through and the **voltage** across $\mathbf{R_L}$ using **Thevenin's theorem. Verify** your answer by **measuring** the voltage and current of the **3Ω** $\mathbf{R_L}$ load.

(b) **Repeat** for a load resistor $\mathbf{R_L}$ **of 9Ω.**

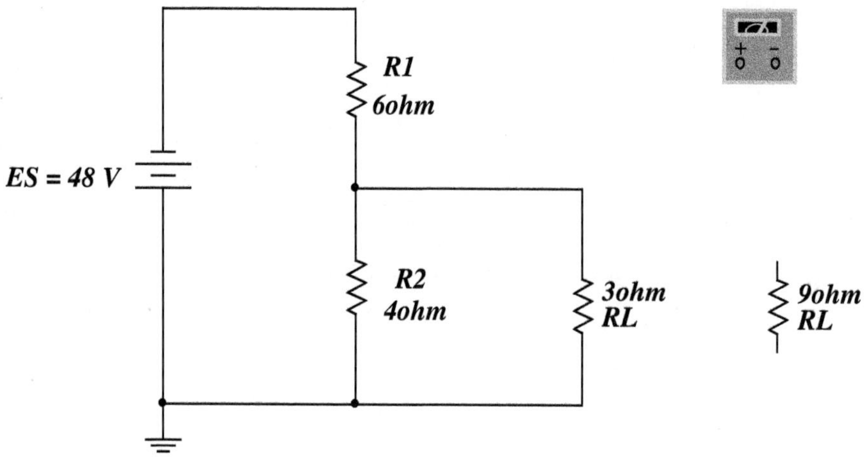

Answers

14-9 Open the Multisim circuit file CH14_09.

(a) Use **Thevenin's theorem** to **determine** and **record** the **current** through and the **voltage** across $\mathbf{R_L}$.

(b) **Verify** your answer by **measuring** and **recording** the **current** flow through and the **voltage** across $\mathbf{R_L}$.

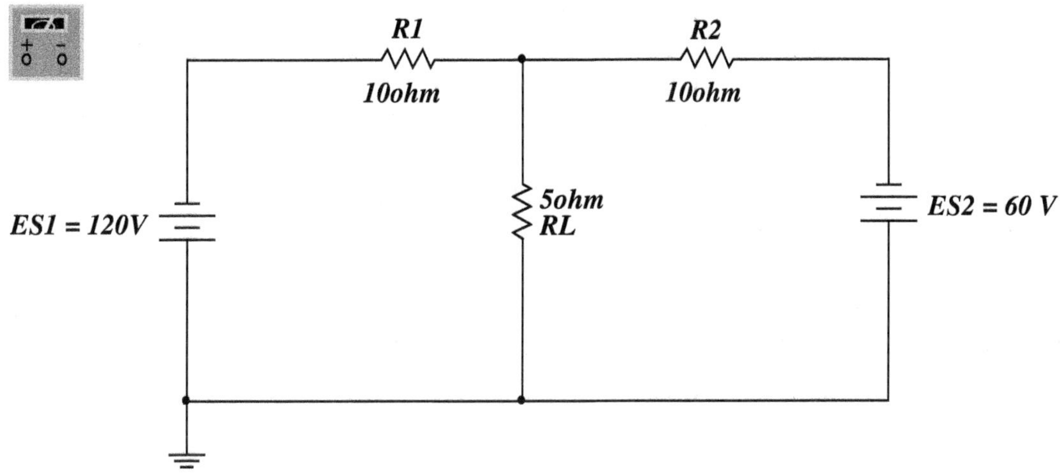

<u>Answers</u>

14-10 Open the Multisim circuit file CH14_10.

(a) Use the **multimeter** to **measure** and **record** the **Thevenin resistance (R_{TH})** and the **Thevenin voltage (E_{TH})** for the circuit.

(b) Use the **Thevenin equivalent circuit** to **determine** and **record** the **current** through and the **voltage** across **R_L**. **Verify** your answer by **measuring** the **current** through and the **voltage** across **R_L**.

(c) **Repeat** part (b) for an **R_L of 5Ω.**

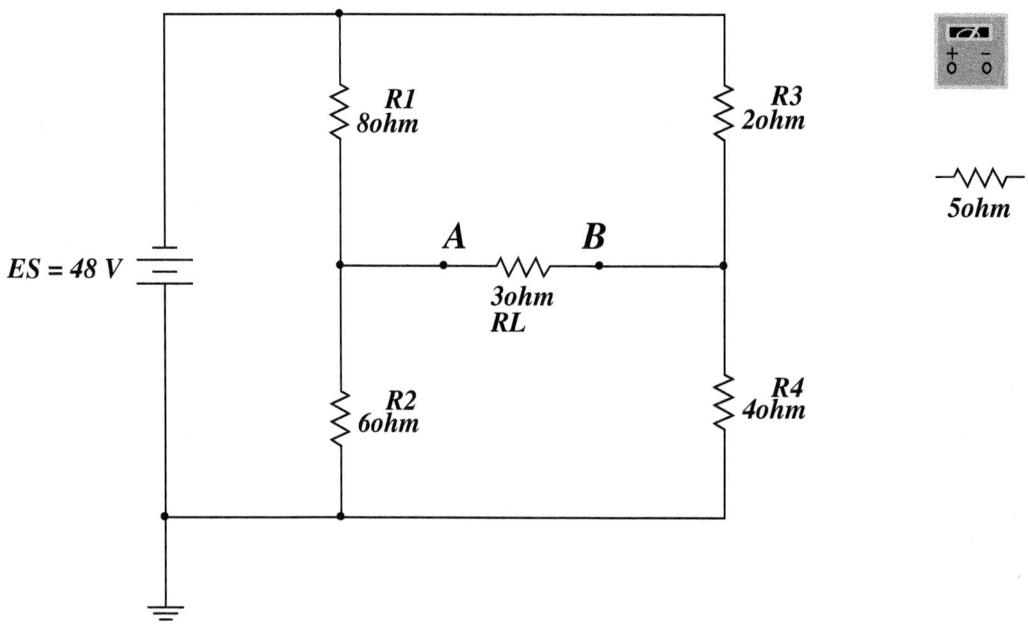

Answers

14-11 Open the Multisim circuit file CH14_11.

(a) Using the **Norton equivalent circuit**, for the circuit shown, **calculate** the **load current** and **voltage** for the **12Ω** load resistor R_L. Turn the simulation on and measure the current through and the voltage across R_L to verify your answer.

(b) Repeat for an **8Ω** load resistor.

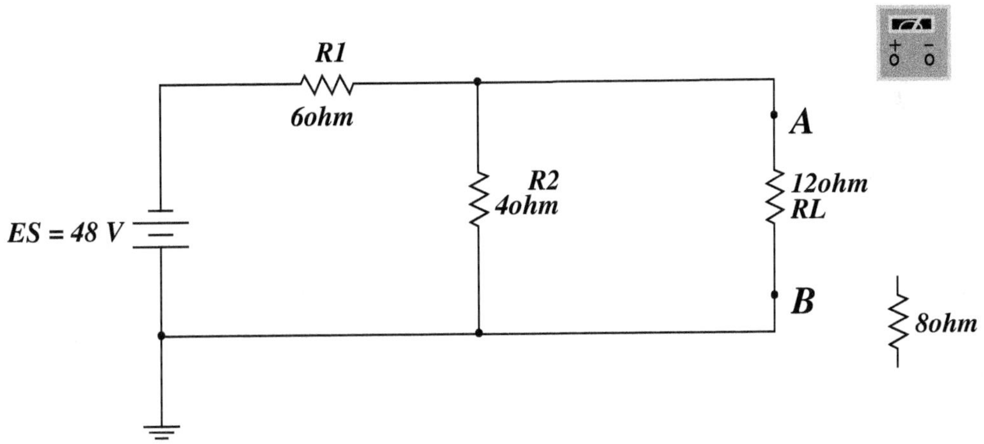

Answers

14-12 Open the Multisim circuit file CH14_12.

(a) Using the **Norton equivalent circuit**, for the circuit shown, **calculate** the **load current** and **voltage** for the **5Ω** load resistor R_L. Turn the simulation on and measure the current through and the voltage across R_L to verify your answer.

(b) Derive the **Thevenin equivalent** for the circuit.

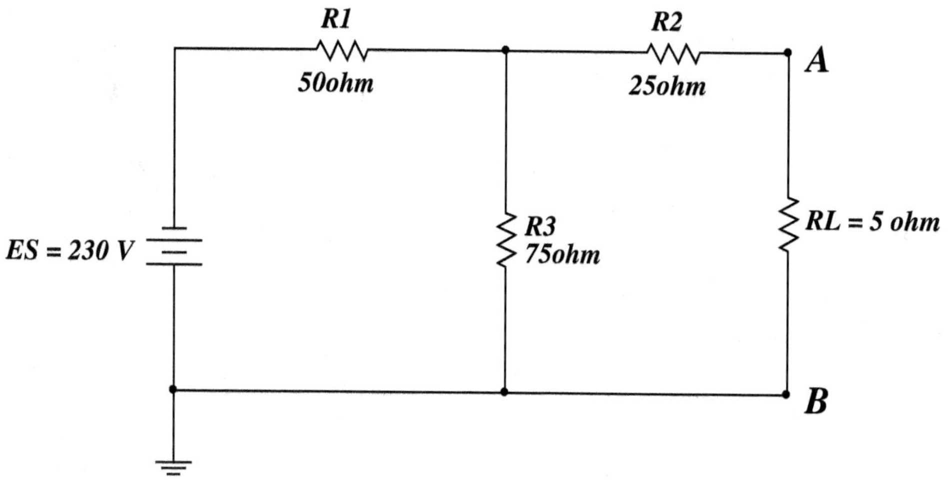

<u>Answers</u>

SECTION 3

Alternating Current (AC) Circuits

Chapter 15	Alternating Current Fundamentals	125
Chapter 16	Inductance and Capacitance	136
Chapter 17	Resistive, Inductive, Capacitive (RLC) Series Circuits	162
Chapter 18	Resistive, Inductive, Capacitive (RLC) Parallel Circuits	186
Chapter 19	Transformers	209

Chapter 15 Alternating Current Fundamentals

15-1 Open the Multisim circuit file CH15_01.

Turn the simulation on and double-click the oscilloscope icon to view the scope display and front panel control screen. According to **the timebase setting of 1 ms/Division,** what is the value of the **time period** for the AC sine waveform displayed?

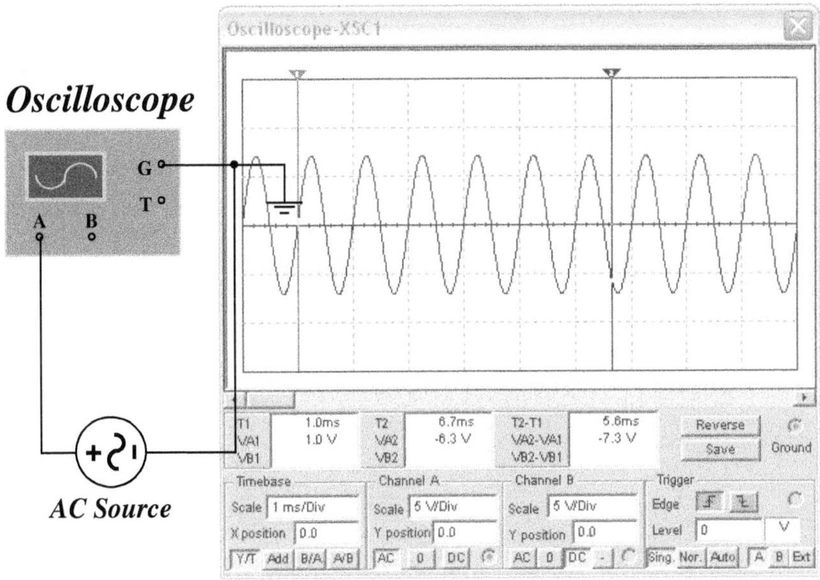

Answer

15-2 Open the Multisim circuit file CH15_02.

Turn the simulation on and double-click the oscilloscope icon to view the scope display and front panel control screen. According to the **5V/Division setting of Channel A**, what is the **peak value** of the AC waveform displayed?

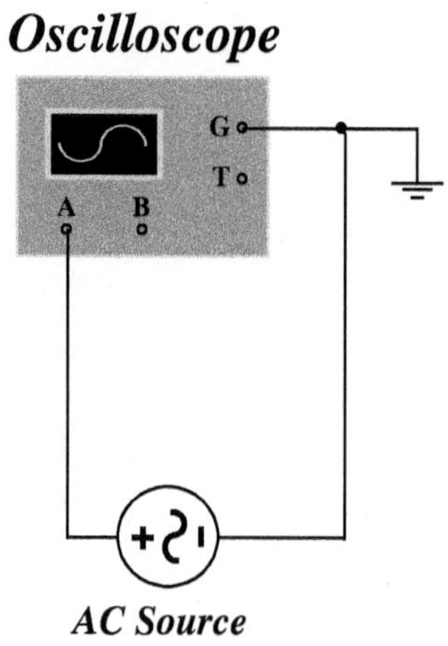

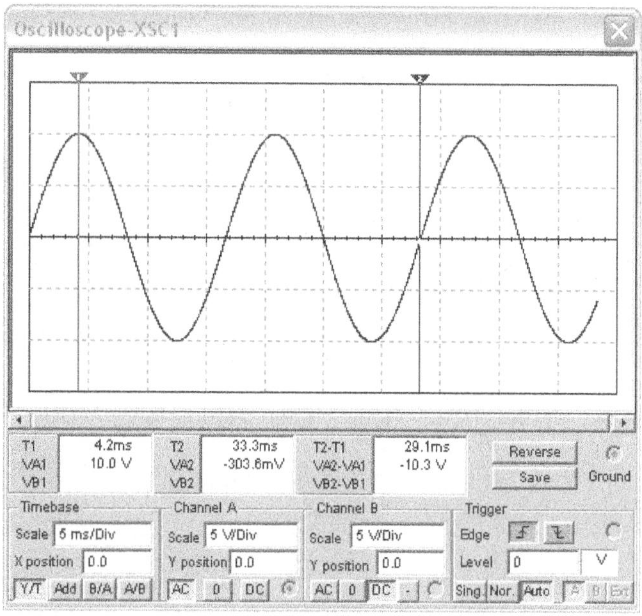

Answer

15-3 Open the Multisim circuit file CH15_03.

Turn the simulation on and double-click the oscilloscope and multimeter icons to view the display and front panel controls.

(a) Which **instrument** measures **rms values**?

(b) What is the **rms value** of the AC voltage?

(c) Which **instrument** measures **peak** and **peak-to-peak** values?

(d) What is the **peak value** of the AC voltage?

(e) What is the **peak-to-peak value** of the AC voltage?

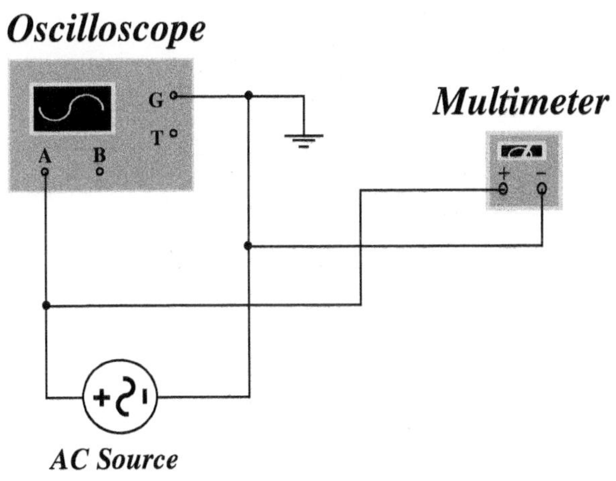

<u>**Answers**</u>

15-4 Open the Multisim circuit file CH15_04.

Turn the simulation on.

(a) What is the peak value of the 120VAC input?

(b) Draw the shape of the DC voltage waveform displayed on the scope.

(c) Calculate the average value of the DC voltage.

(d) Connect the multimeter and record the value of the DC output voltage.

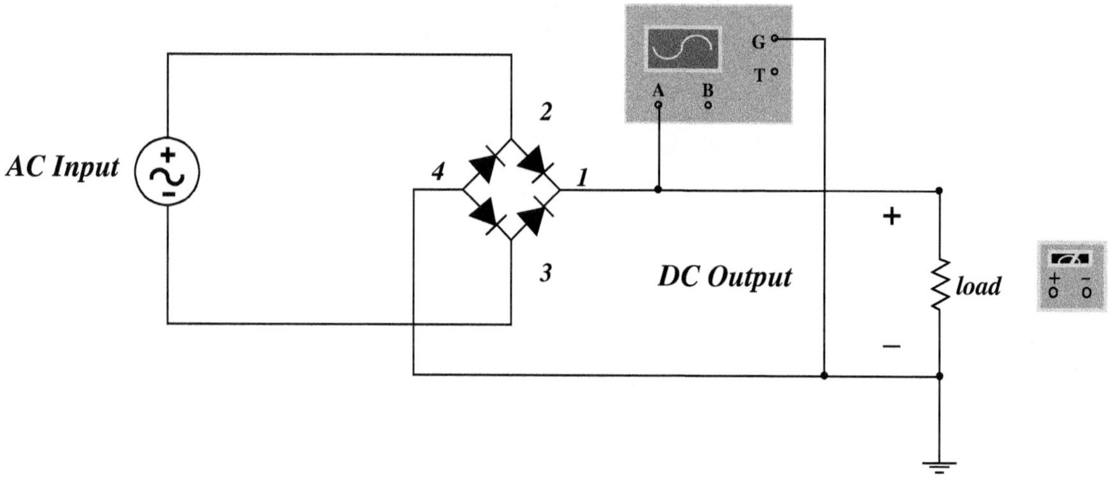

<u>Answers</u>

15-5 Open the Multisim circuit file CH15_05.

(a) Record the value of the high-voltage transformer primary.

(b) Measure and record the voltage from the Black (Hot) to the White (Neutral).

(c) Measure and record the voltage from the Red (Hot) wire to the White (Neutral).

(d) Measure and record the voltage from the Black (Hot) to the Red (Hot).

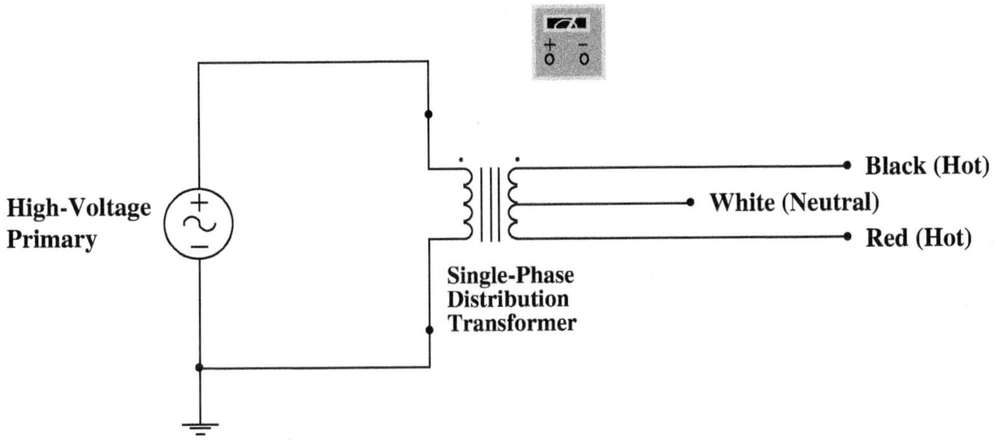

Answers

15-6 Open the Multisim circuit file CH15_06.

For the **Three Phase Y** connected source:

(a) Exam the Phase A and Phase B waveforms displayed on the scope. How many electrical **degrees separate Phase B from Phase A?**

(b) Use the **multimeter** to **record** the value of the voltage from **Phase A to Neutral.**

(c) **Calculate** the line voltage from **Phase A to Phase B.**

(d) **Measure** and **record** the value of line voltage from **Phase A to Phase B.**

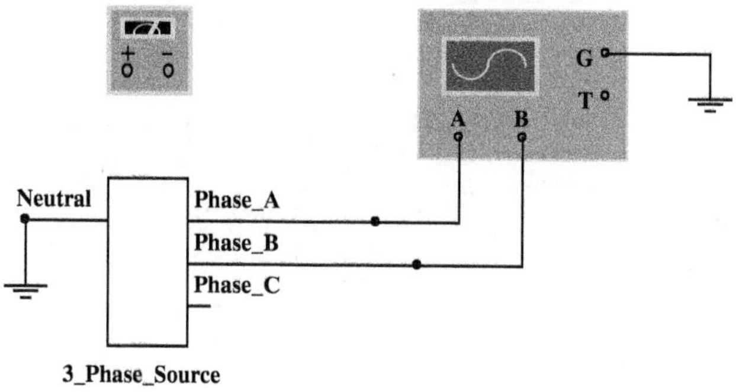

Y Connected

<u>**Answers**</u>

15-7 Open the Multisim circuit file CH15_07.

For the **Three Phase Y** connected source:

(a) Use the **multimeter** to **record** the value of the line voltage from **A-to-B**, **A-to-C**, and **B-to-C.**

(b) **Calculate** the phase voltage from line **A, B, or C to Neutral**.

(c) **Measure** and **record** the phase voltage from line **A, B, or C to Neutral**.

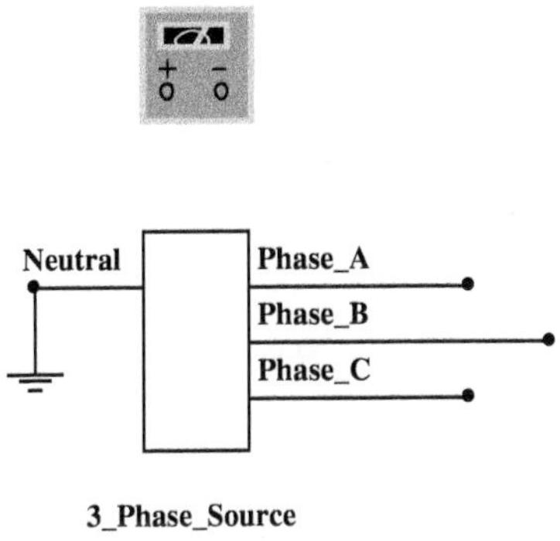

Y Connected

Answers

15-8 Open the Multisim circuit file CH15_08.

(a) Determine all values of **E, I, and R** for the **AC resistive series circuit** and **record** your findings in the table provided.

(b) **Measure** all values of **E, I, R** and record your measurements in the table provided.

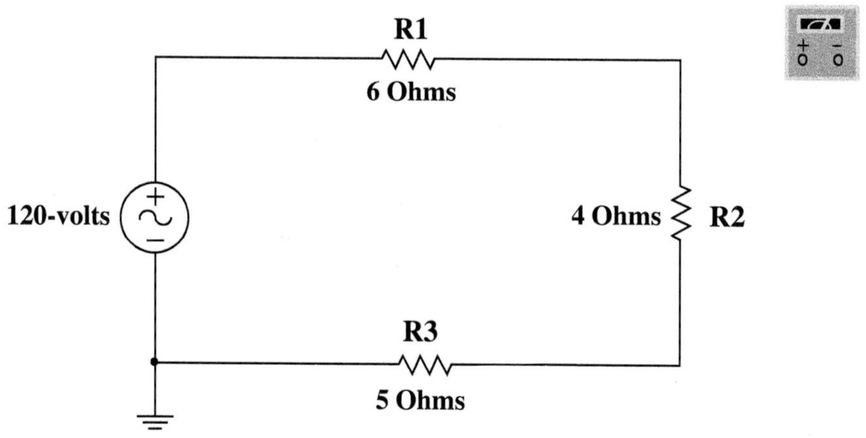

<u>**Answers**</u>

(a)

	Voltage	**Current**	**Resistance**
R_1			
R_2			
R_3			
Total			

(b)

	Voltage	**Current**	**Resistance**
R_1			
R_2			
R_3			
Total			

15-9 Open the Multisim circuit file CH15_09.

(a) **Calculate** and **record** the **effective total power** consumed by the circuit.

(b) Turn the simulation on and **record the total power** consumed as indicated by the wattmeter reading.

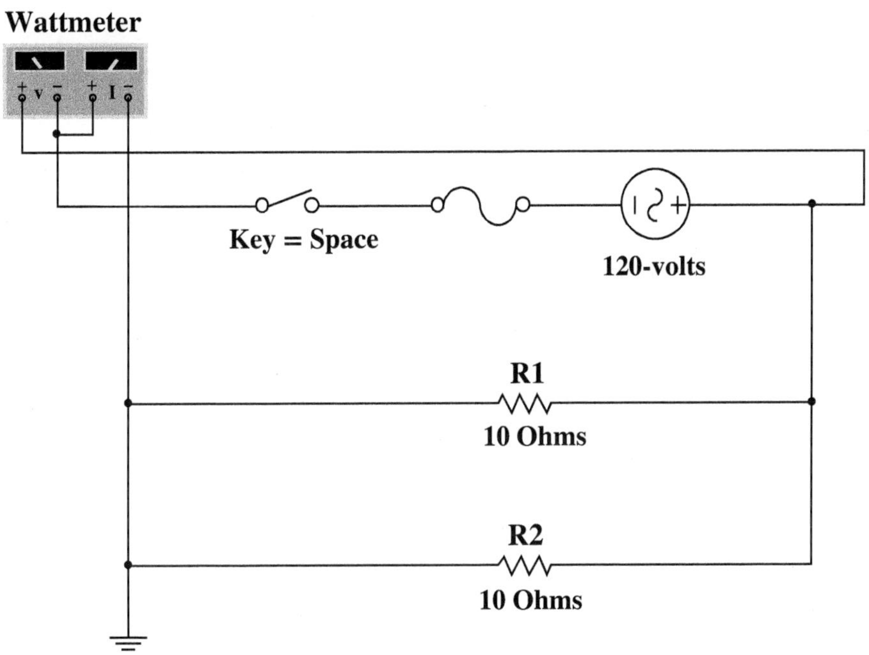

Answers

15-10 Open the Multisim circuit file CH15_10.

(a) Connect the **multimeter** and **record** the value of the **line current and voltage.**

(b) Connect the **multimeter** and **record** the value of the **phase current and voltage.**

(c) **Calculate** the **total power** based on the **line current and voltage readings.**

(d) **Calculate** the total power based on the **phase current and voltage readings.**

(e) **Record** the amount of **power consumed by one phase** as indicated by the wattmeter reading.

(f) How is the **total power** determined **from the wattmeter reading?**

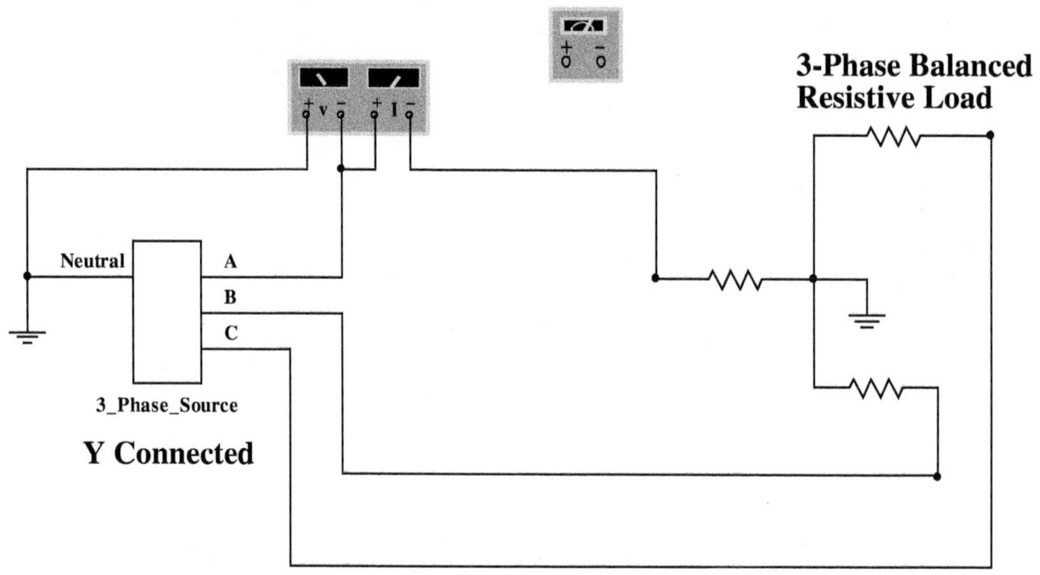

<u>**Answers**</u>

15-11 Open the Multisim circuit file CH21_12.

Turn the simulation on.

(a) **Record** the value of the line **current** and line **voltage.**

(b) **Calculate** the **total power** based on these current and voltage readings.

(c) Double-click the wattmeter icons and **record** the readings $\mathbf{W_1}$ and $\mathbf{W_2}$.

(d) How is the **total power determined** from the two **wattmeter readings?**

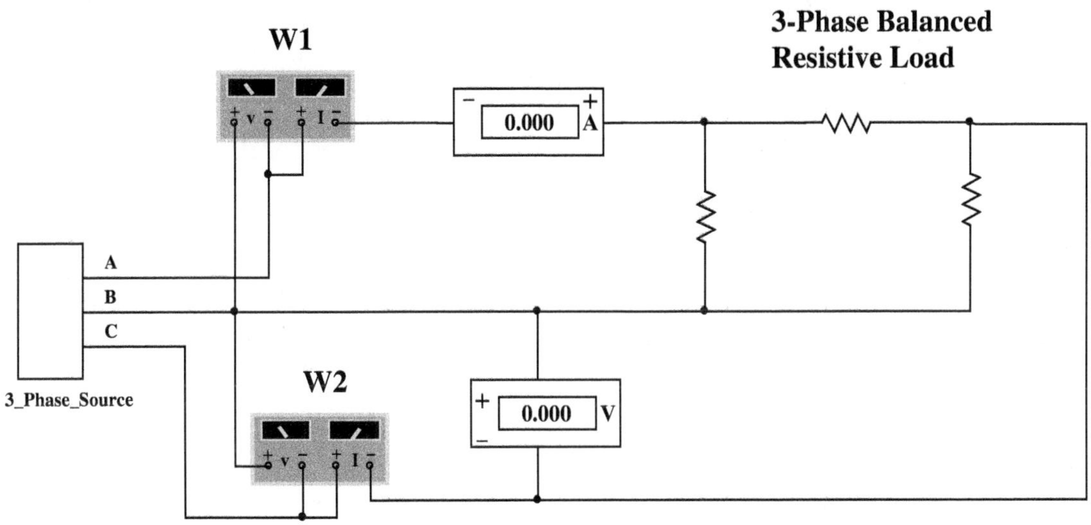

<u>Answers</u>

Chapter 16 Inductance and Capacitance

16-1 Open the Multisim circuit file CH16_01.

(a) **Calculate** the value of the **time constant**.

(b) Turn the simulation on, and operate the switch to open and close the circuit to the coil. **Sketch the voltage waveform**, as displayed by the scope, for the circuit on first closing and opening.

(c) **Calculate** the **time constant** when the value of the inductance is **10 H** and the value of the resistance remains at **100 Ω**.

(d) Double-click on the inductor and change its inductance value to 10 H. Operate the simulation and make **note** of the **relative time period** it takes for the inductor to reach its maximum and zero values. **Explain** the **reason** for this change in time.

(e) **Change** the **inductance value to zero** to create a purely resistive circuit. Operate the simulation and make **note** of the **relative time period** it takes for the inductor to reach its maximum and zero values. **Explain** the **reason** for this change in time.

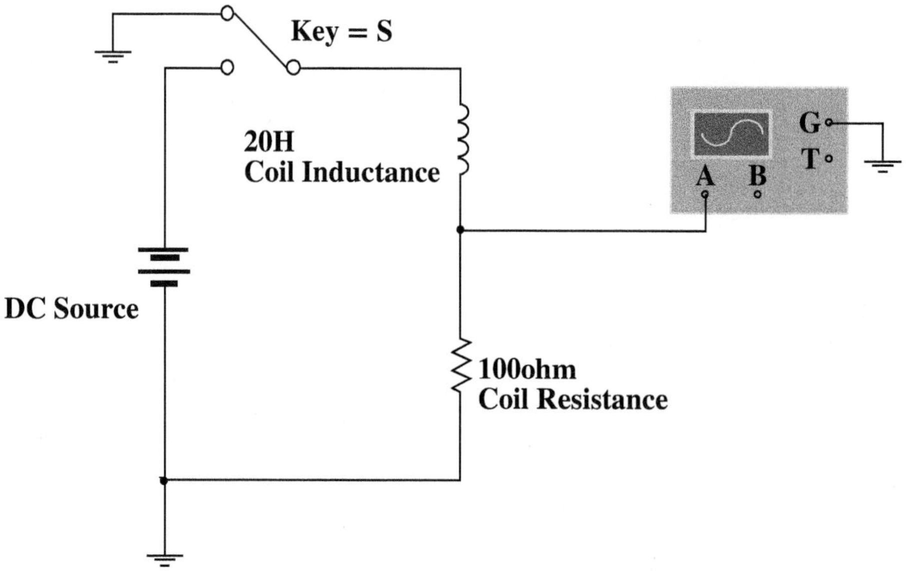

Answers

16-2 Open the Multisim circuit file CH16_02.

Turn the simulation on.

(a) **Record** the value of the **current** with the **12 VDC** source powering the circuit.

(b) **Record** the value of the **current** with the **12 VAC** source powering the circuit.

(c) What type of load (**resistive - inductive - capacitive**) is the **lamp classified** as?

(d) What type of load (**resistive - inductive - capacitive**) is the **coil classified** as?

(e) **Which power source** produces the **greatest** amount of **current**? Why?

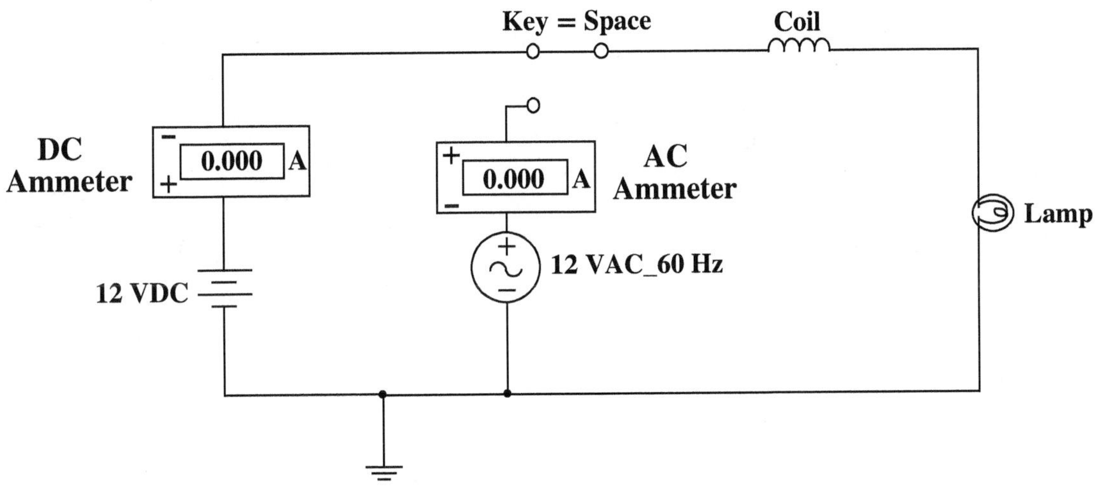

<u>Answers</u>

16-3 Open the Multisim circuit file CH16_03.

Turn the simulation on.

(a) **Record** the value of the **current** with the **5 mH** coil in series with the lamp.

(b) **Record** the value of the **current** with the **10 mH** coil in series with the lamp.

(c) **Which inductor** connection produces the **greatest** amount of **current**? **Why**?

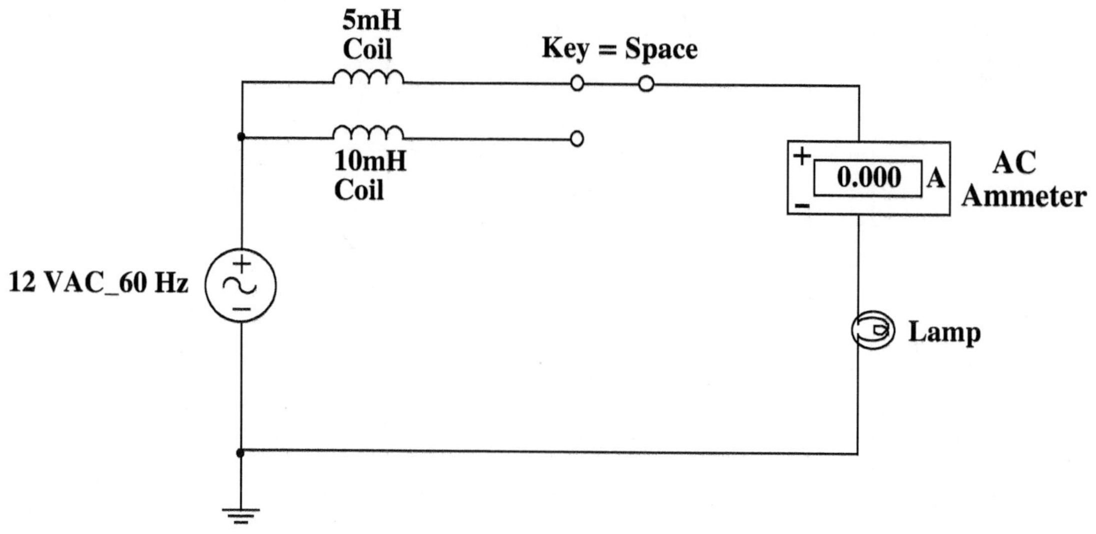

<u>Answers</u>

16-4 Open the Multisim circuit file CH16_04.

Turn the simulation on.

(a) **Record** the value of the **current** with the **60 Hz** source powering the circuit.

(b) **Record** the value of the **current** with the **300 Hz** source powering the circuit.

(c) **Which power** source produces the **greatest** amount of **current**?

Why?

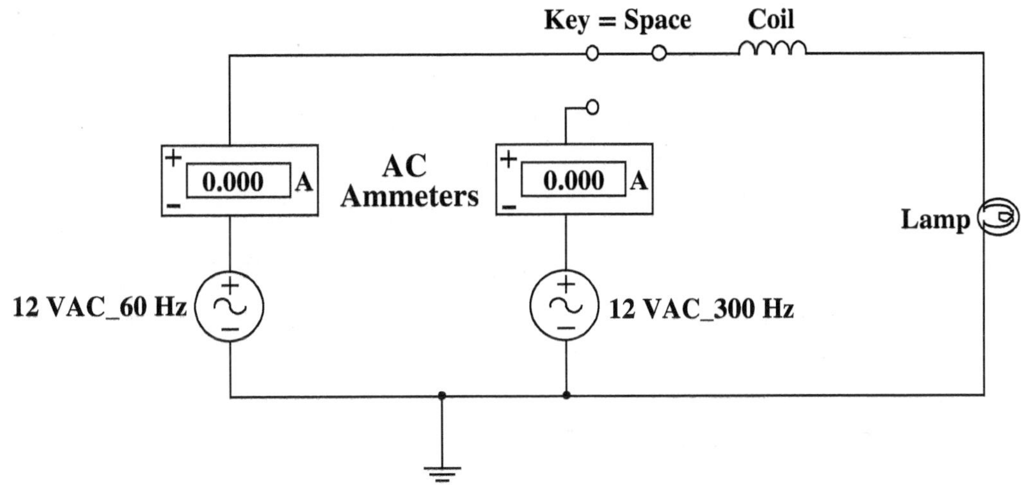

Answers

16-5 Open the Multisim circuit file CH16_05.

Turn the simulation on.

(a) Connect the multimeter to **measure** and **record** the **DC resistance** of the **ideal or pure** inductor **L**.

(b) **Calculate** the **inductive reactance** **(X_L)** of inductor **L**.

(c) **Calculate** the value of the circuit **current** flow.

(d) Connect the multimeter to **measure** and **record** the value of the **current** flow.

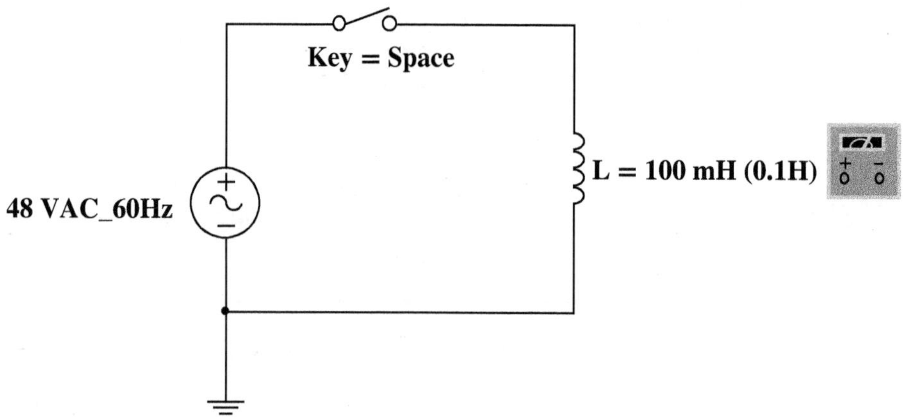

<u>**Answers**</u>

16-6 Open the Multisim circuit file CH16_06.

(a) **Calculate** the value of the total **circuit inductance (L_T)**.

(b) **Calculate** the value of the total **circuit inductive reactance (X_L)**.

(c) **Calculate** the value of the circuit **current** flow.

(d) Connect the multimeter to **measure** and **record** the **current** flow.

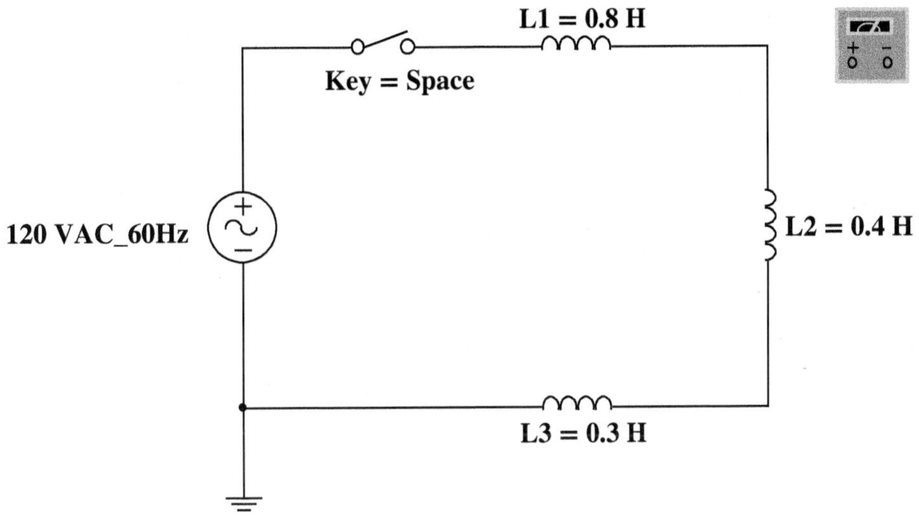

Answers

16-7 Open the Multisim circuit file CH16_07.

(a) **Calculate** the value of the **total** circuit **inductance (L_T)**.

(b) **Calculate** the value of the **total** circuit **inductive reactance (X_L).**

(c) **Calculate** the value of the **total circuit current** flow.

(d) Connect the multimeter to **measure** and **record** the **total current** flow.

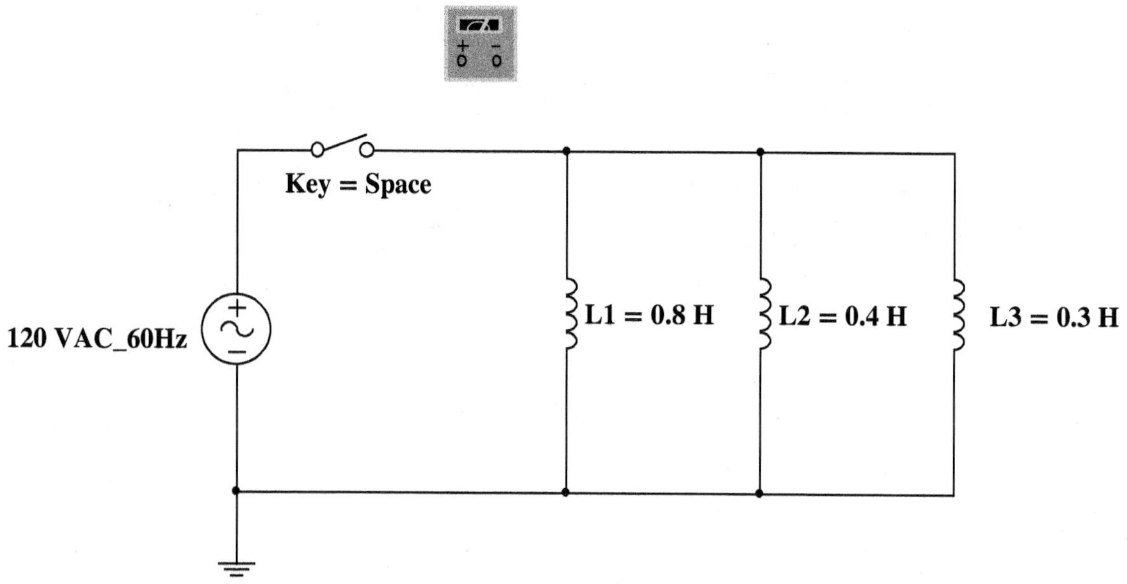

<u>**Answers**</u>

16-8 Open the Multisim circuit file CH16_08.

This simulation is designed to show the **current and voltage phase relationship** for **pure resistive** and **pure inductive** loads. The red trace of the scope represents the voltage and the blue trace represents the current.

Turn the simulation on and **record** the following observations:

(a) Phase **relationship (in-phase, lags, or leads)** between **current and voltage** with the switch set to the **resistor position.**

(b) Phase **relationship (in-phase, lags, or leads)** between **current and voltage** with the switch set to the **inductor position.**

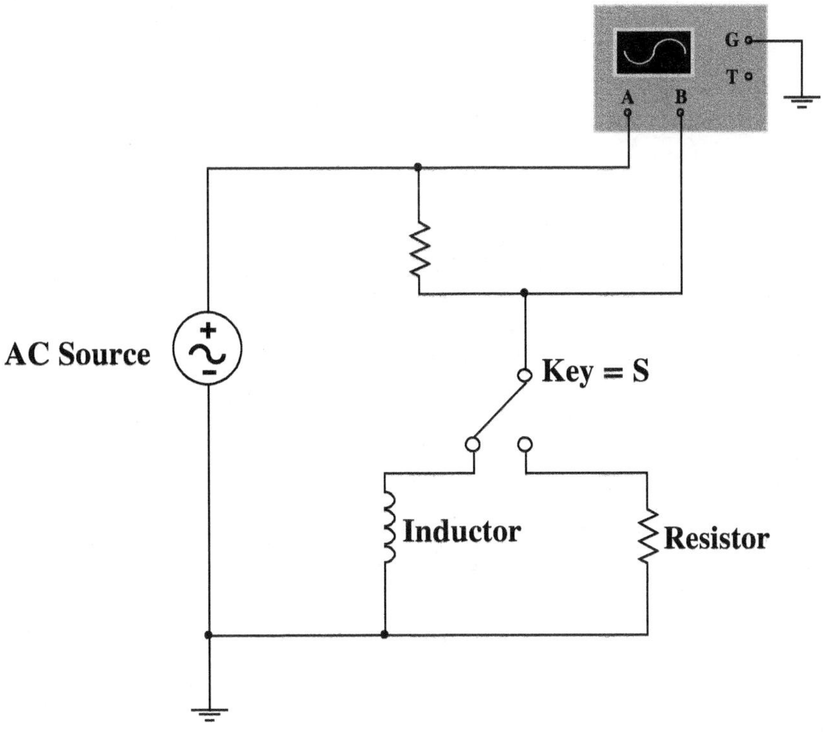

<u>Answers</u>

16-9 Open the Multisim circuit file CH16_09.

Turn the simulation on.

(a) **Record** the value of the **voltage, current, and wattage** of the circuit with the **resistive** load connected.

(b) **Record** the value of the **voltage, current, and wattage** of the circuit with the **inductive** load connected.

(c) **Which measurement** is **noticeably different** for the two components? **Why?**

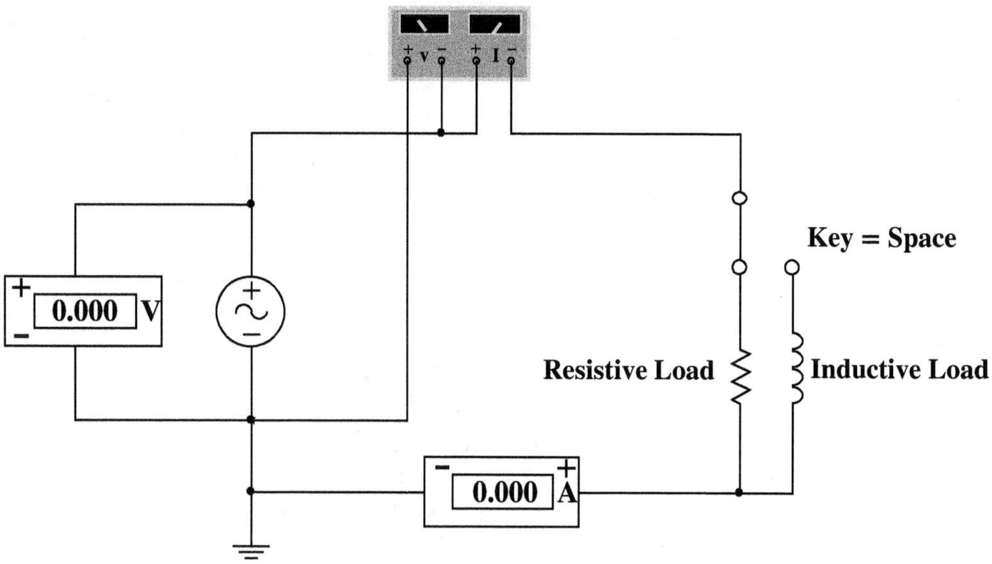

<u>**Answers**</u>

16-10 Open the Multisim circuit file CH16_10.

(a) **Calculate** the value of the **inductive reactance (X_L).**

(b) **Calculate** the value of the **current** flow.

(c) Connect the multimeter to **measure** and **record** the value of the **current** flow.

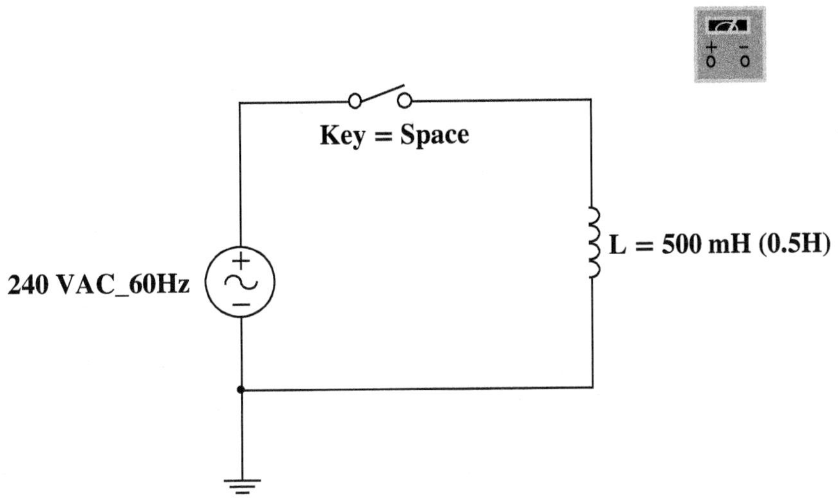

<u>Answers</u>

16-11 Open the Multisim circuit file CH16_11.

(a) **Calculate** the value of the **total inductance (L_T).**

(b) **Calculate** the value of the total **inductive reactance (X_L total).**

(c) **Calculate** the value of the **current** flow.

(d) Connect the multimeter to **measure** and **record** the value of the **current** flow.

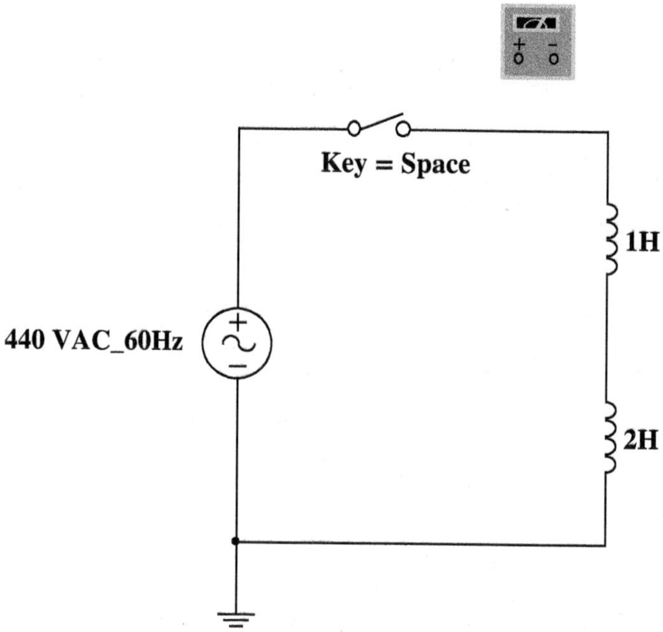

Answers

16-12 Open the Multisim circuit file CH16_12.

(a) **Calculate** the value of the **total inductance (L_T).**

(b) **Calculate** the value of the total **inductive reactance (X_L total).**

(c) **Calculate** the value of the **current** flow.

(d) Connect the multimeter to **measure** and **record** the value of the **current** flow.

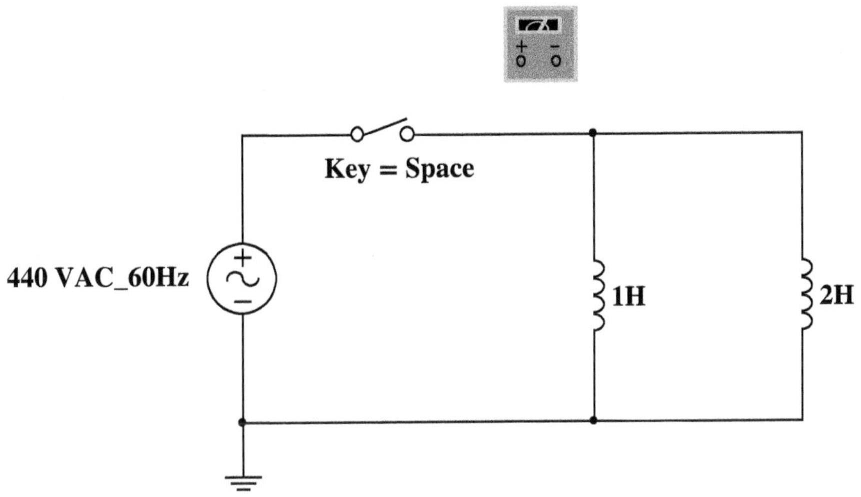

<u>**Answers**</u>

16-13 Open the Multisim circuit file CH16_13.

(a) **Calculate** the **inductive reactance (X_L)** of the inductive load.

(b) **Calculate** what the **ammeter reading** should be.

(c) **Determine** what the **wattmeter reading** should be.

(d) Turn the simulation on and **record** the value of the **ammeter and wattmeter readings.**

(e) **Calculate** the value of the **Volt-Amps Reactive (VARs)** for the circuit.

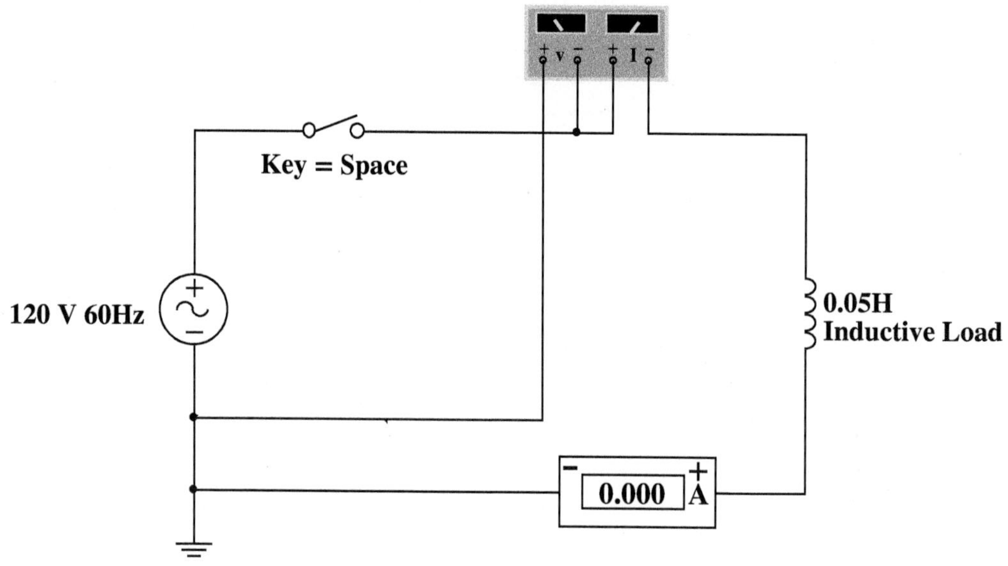

<u>Answers</u>

16-14 Open the Multisim circuit file CH16_14.

Turn the simulation on.

(a) Operate the switch to **charge and discharge** the **capacitor**. According to the oscilloscope waveform, what amount of time **(long, short, or instantaneous)** does it takes for the capacitor to fully **charge** and **discharge**?

(b) **Repeat** with the **1 kΩ resistor connected in series** with the capacitor.

(c) **Repeat** with the **10 kΩ resistor connected in series** with the capacitor.

(d) How does the **resistance value** of the series resistor affect the time it takes for the capacitor to fully **charge** and **discharge**?

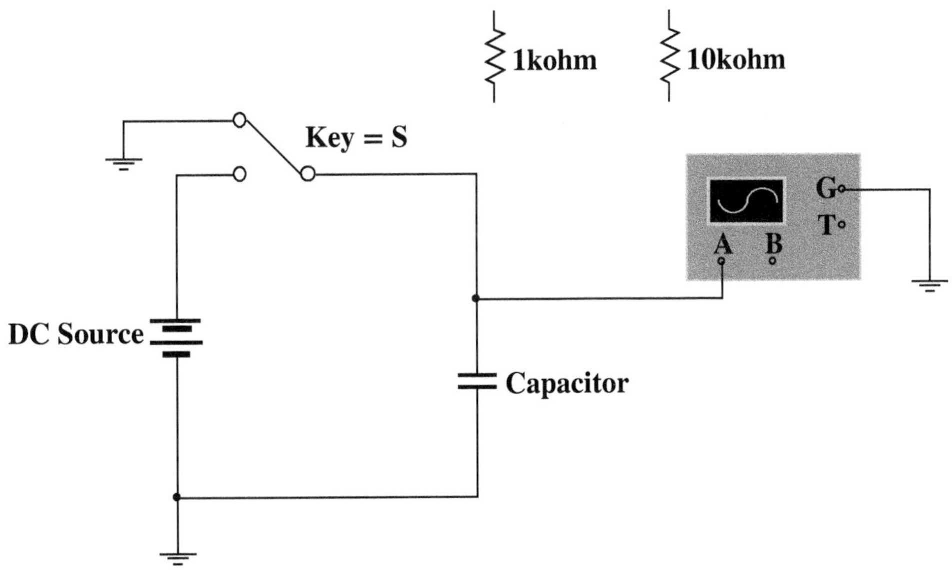

Answers

16-15 Open the Multisim circuit file CH16_15.

(a) **Calculate** the value of the circuit **RC time constant**.

(b) Approximately **how long** will it take the capacitor to **fully charge and discharge**? Turn the simulation on and operate the circuit to verify your answer.

(c) **Calculate** the value of the **RC time constant** with the resistor value changed to **50 kΩ.**

(d) Approximately **how long** will it take the capacitor to **fully charge and discharge** using the **50 kΩ resistor**? Turn the simulation on and operate the circuit to verify your answer.

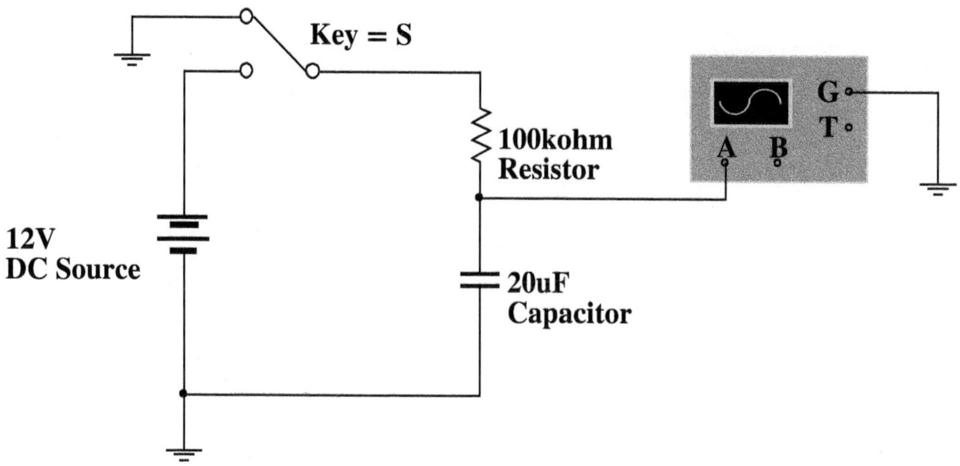

Answers

16-16 Open the Multisim circuit file CH16_16.

Turn the simulation on.

(a) Apply 6 volts **DC** and record the state **(ON or OFF)** of the lamp.

(b) Apply 6 volts **AC** and record the state **(ON or OFF)** of the lamp.

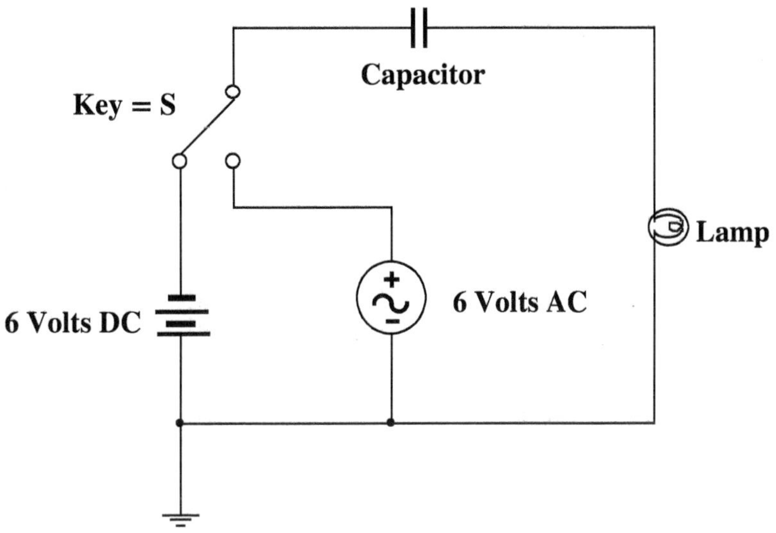

<u>**Answer**</u>

16-17 Open the Multisim circuit file CH16_17.

Turn the simulation on.

(a) **Record** the amount of **curren**t flow with the **10µF capacitor** connected in series with the lamp.

(b) **Record** the amount of **current** flow with the **50µF capacitor** connected in series with the lamp.

(c) Which **capacitor** connection resulted in the **most current flow**?

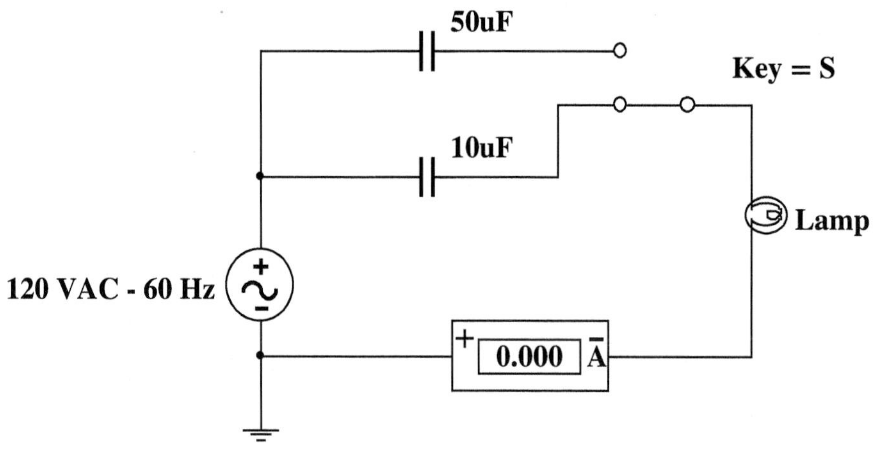

Answers

16-18 Open the Multisim circuit file CH16_18.

Turn the simulation on.

(a) **Record** the amount of **current** flow with the **120 V, 60 Hz** power supply connected in series with the lamp.

(b) **Record** the amount of **current** flow with the **120 V, 600 Hz** power supply connected in series with the lamp.

(c) Which **frequency** resulted in the **most current flow**?

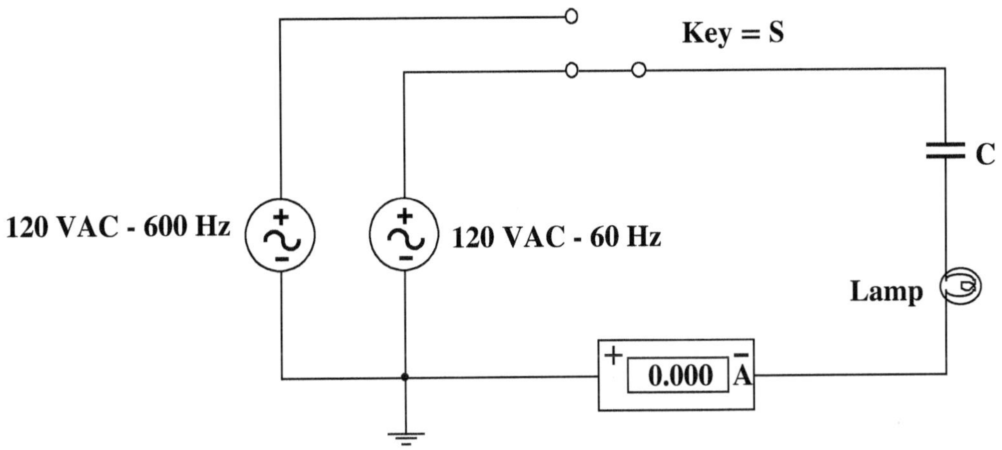

<u>Answers</u>

16-19 Open the Multisim circuit file CH16_19.

(a) **Calculate** the **capacitive reactance** of the capacitor.

(b) **Calculate** the value of the **current** flow in the circuit.

(c) **Measure** and **record** the amount of **current** flow in the circuit.

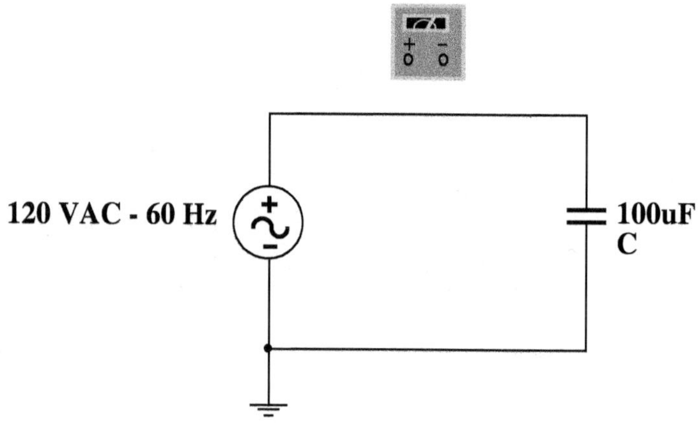

<u>Answers</u>

16-20 Open the Multisim circuit file CH16_20.

(a) **Calculate** the **total capacitance** of the circuit.

(b) **Calculate** the **total capacitive reactance** of the circuit.

(c) **Calculate** the value of the **current** flow in the circuit.

(d) **Measure** and **record** the amount of **current** flow in the circuit.

(e) **Calculate** the value of the **voltage** drop **across each capacitor**.

(f) **Measure** and **record** the value of the **voltage** drop **across each capacitor.**

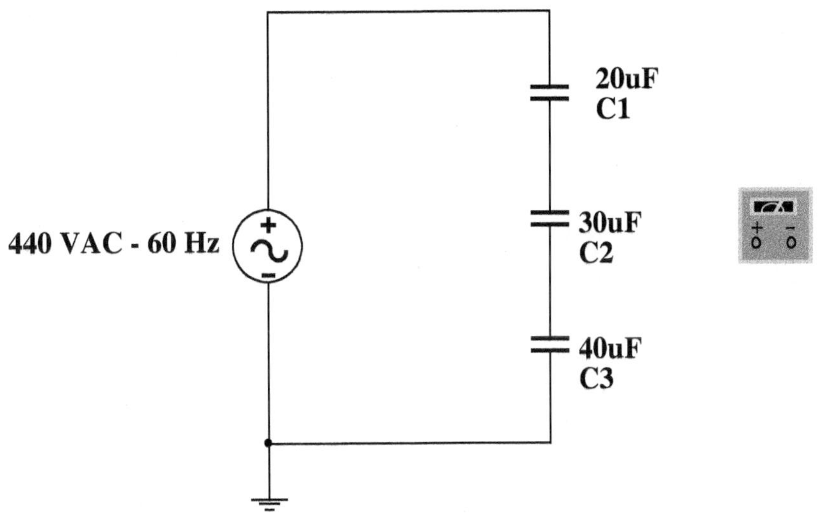

<u>**Answers**</u>

16-21 Open the Multisim circuit file CH16_21.

(a) **Calculate** the **total capacitance** of the circuit.

(b) **Calculate** the **total capacitive reactance** of the circuit.

(c) **Calculate** the value of the total **current** flow of the circuit.

(d) **Measure** and **record** the amount of **total current** flow of the circuit.

(e) **Calculate** the value of the **current reactance** of **each capacitor**.

(f) **Calculate** the value of the **current** flow **through each capacitor**.

(g) **Measure** and **record** the value of the **current** flow **through each capacitor.**

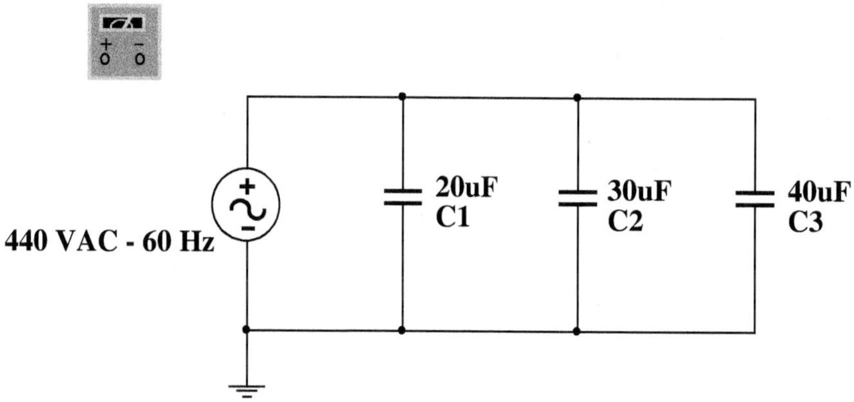

Answers

16-22 Open the Multisim circuit file CH16_22.

This simulation is designed to show the current and voltage phase relationship for a pure capacitive load. The red trace of the scope represents the voltage and the blue trace represents the current. Turn the simulation on and study the current and voltage waveforms relative to each other. What is the **phase relationship** between the **current** and the **voltage**?

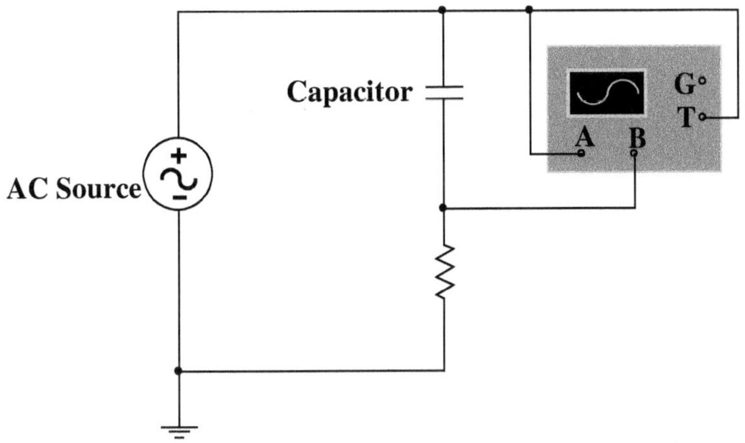

Answer

16-23 Open the Multisim circuit file CH16_23.

Turn the simulation on.

(a) **Record** the value of the **current** flow through the **inductor** and **capacitor.**

(b) **Calculate** the value of the **total line current** based on the fact that these two current are **180° out-of-phase** and act to oppose each other.

(c) **Measure** and **record** the value of the **line current**.

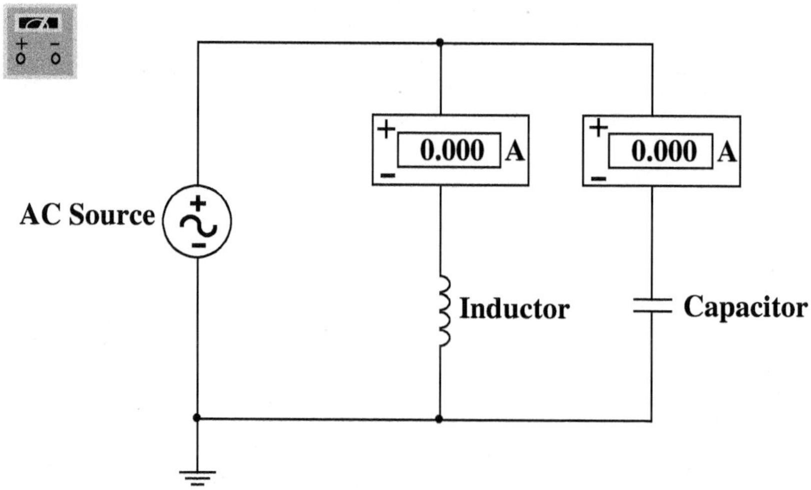

Answers

16-24 Open the Multisim circuit file CH16_24.

Turn the simulation on.

(a) **Record** the value of the **voltage, current, and wattage** of the circuit with the **resistive** load connected.

(b) **Record** the value of the **voltage, current, and wattage** of the circuit with the **inductive** load connected.

(c) Which measurement is **noticeably different** for the two components? **Why?**

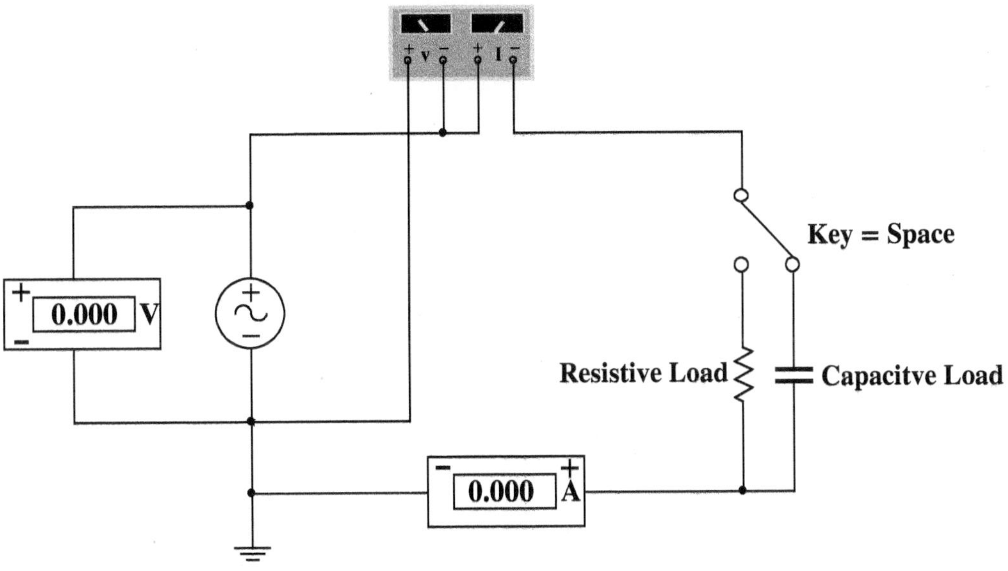

<u>Answers</u>

16-25 Open the Multisim circuit file CH16_25.

(a) **Calculate** the **capacitance reactance** of the capacitor.

(b) **Calculate** the value of the **current** flow.

(c) **Measure** and **record** the value of the **current** flow.

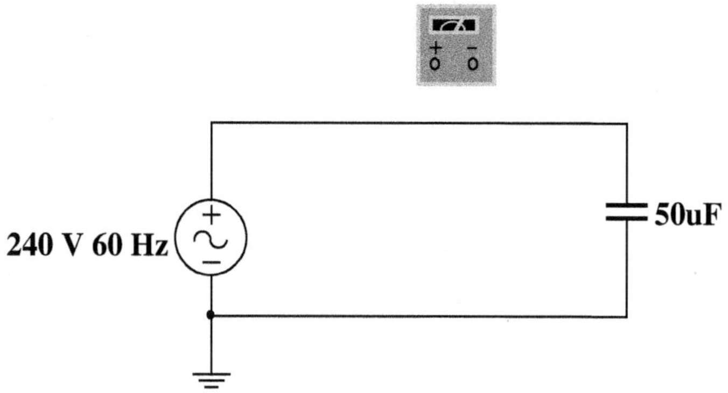

__Answers__

16-26 Open the Multisim circuit file CH16_26.

(a) **Calculate** the value of the **capacitive reactance.**

(b) **Calculate** the value of the **current**.

(c) **Calculate** the value of the **Volt-Amps Reactive (VARs).**

(d) **Determine** the value of the **wattmeter reading.**

(e) Turn the simulation on and **record** the value of the **ammeter** and **wattmeter** readings.

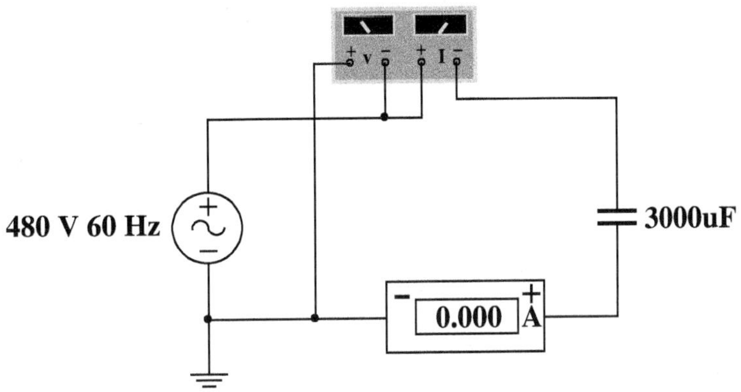

Answers

Chapter 17 Resistive, Inductive, Capacitive (RLC) Series Circuits

17-1 Open the Multisim circuit file CH17_01.

(a) Turn the simulation on and observe voltage A and B.

(b) What is the phase relationship **(in-phase or out-of-phase)** of the two voltages?

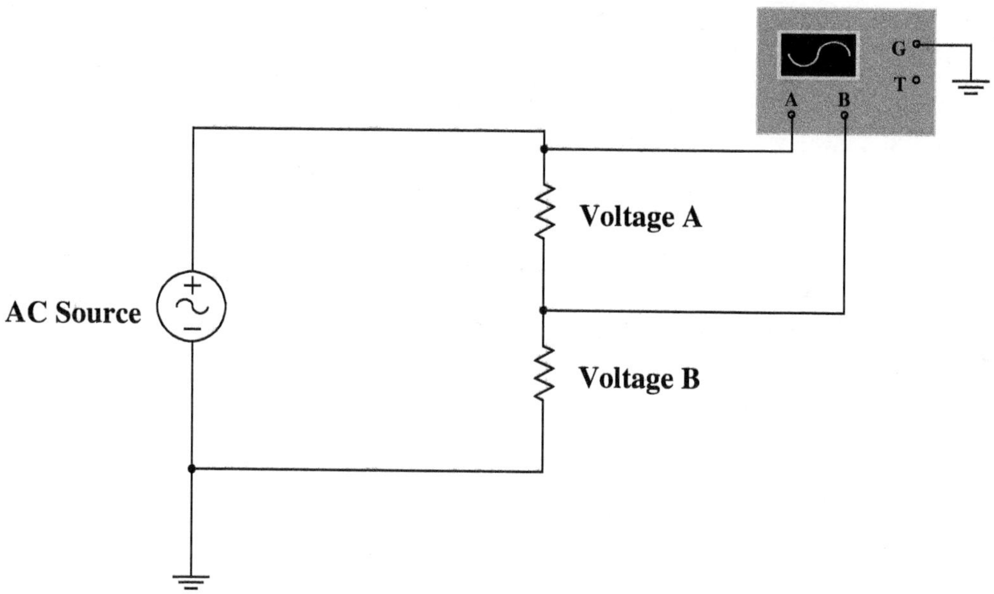

Answer

17-2 Open the Multisim circuit file CH17_02.

(a) Turn the simulation on and observe voltage A and B.

(b) What is the phase relationship **(A leads B by 90° or B leads A by 90°)** of the two voltages?

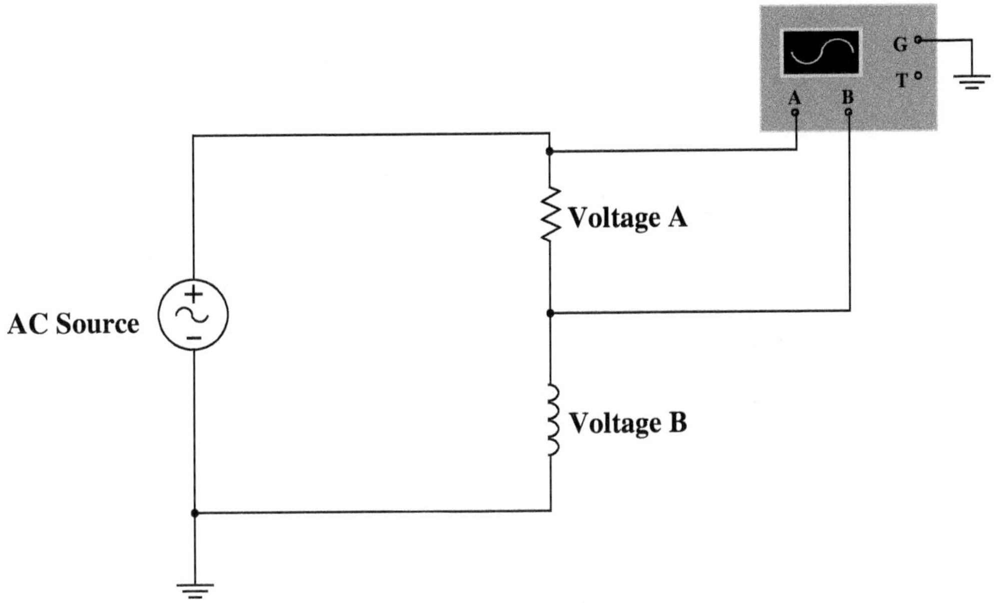

<u>**Answer**</u>

17-3 Open the Multisim circuit file CH17_03.

(a) **Calculate** the output **voltage** across the **load** for circuit **# 1**.

(b) **Calculate** the output **voltage** across the **load** for circuit **# 2.**

(c) **Measure** and **record** the **voltage** across the **load** for circuit **# 1**.

(d) **Measure** and **record** the **voltage** across the **load** for circuit **# 2**.

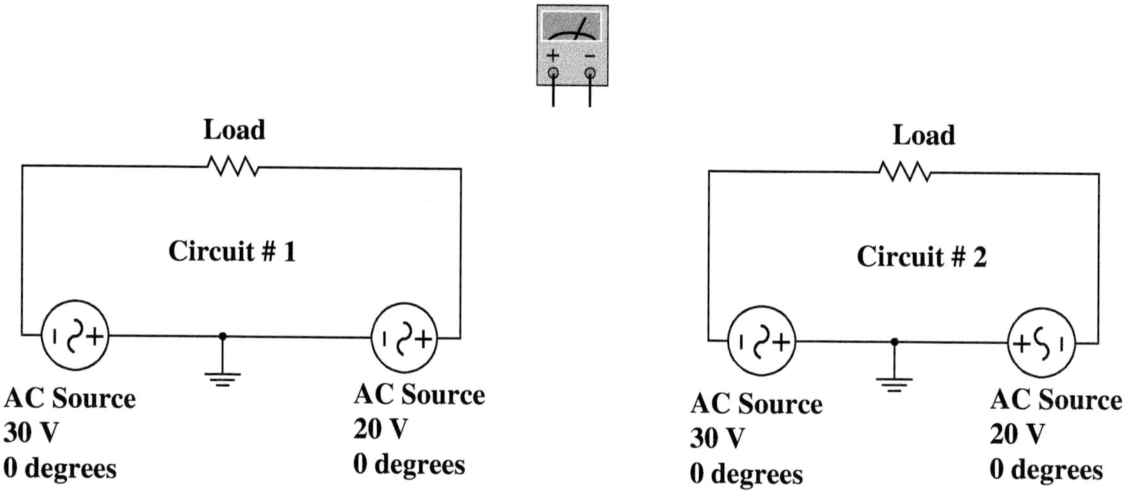

<u>**Answers**</u>

17-4 Open the Multisim circuit file CH17_04.

Measure and **record** the output **voltage** across the **load**.

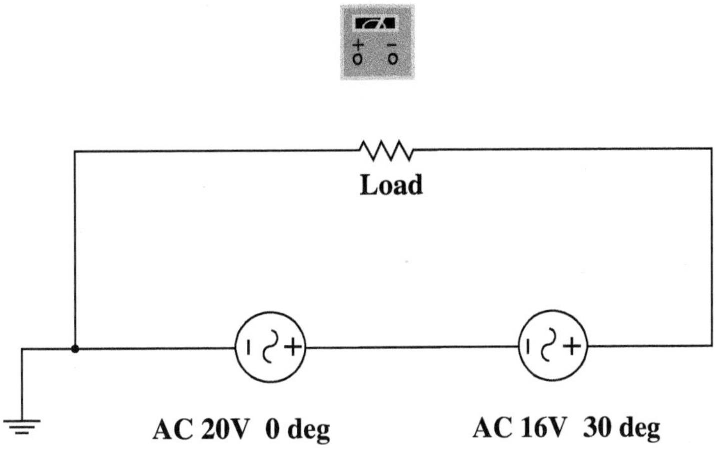

Answers

17-5 Open the Multisim circuit file CH17_05.

Measure and **record** the output **voltage** across the **load**.

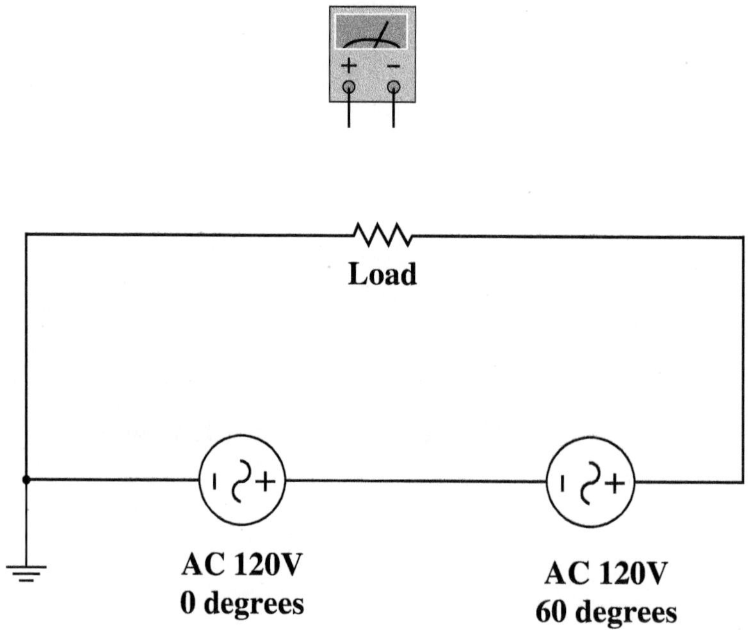

<u>Answers</u>

17-6 Open the Multisim circuit file CH17_06.

(a) **Calculate** the output **voltage** across the **load** for the circuit shown.

(b) **Measure** and **record** the output **voltage** across the **load**.

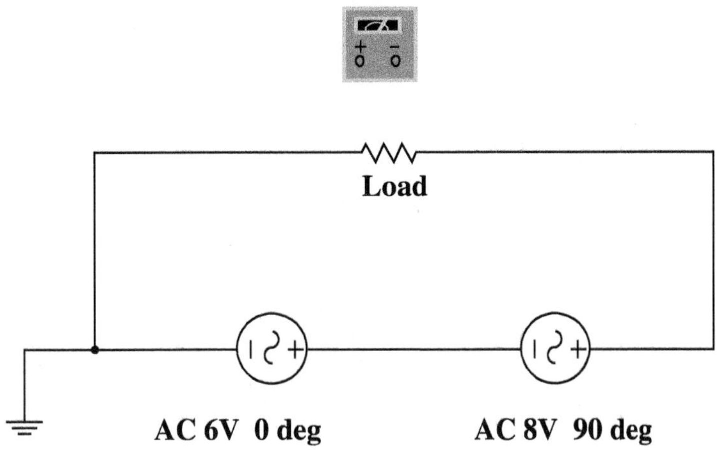

<u>Answers</u>

17-7 Open the Multisim circuit file CH17_07.

(a) **Calculate** the **impedance (Z)** for the circuit.

(b) **Calculate** the **current** flow for the circuit.

(c) **Measure** and **record** the value of the **current** flow.

(d) **Calculate** the **phase angle theta (θ)** for the circuit.

Inputs A and B of the oscilloscope are connected to show the ∠θ for the circuit. **Examine** the phase shift between the two waveforms displayed.

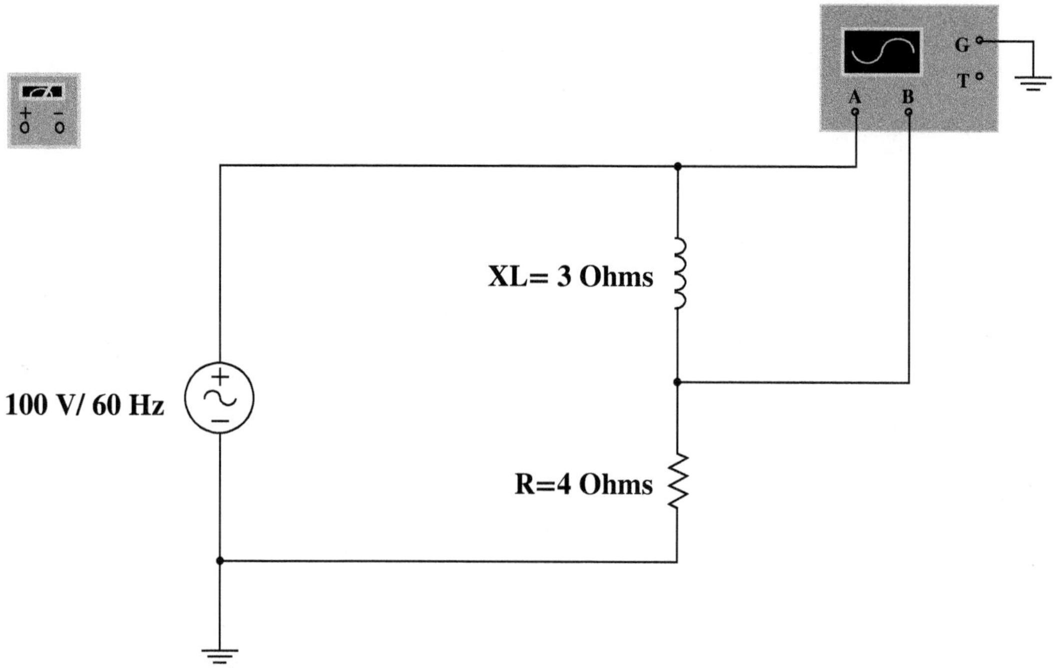

Answers

17-8 Open the Multisim circuit file CH17_08.

(a) **Calculate** the **current** flow for the circuit.

(b) **Measure** and **record** the **current** flow.

(c) **Calculate** the **voltage** across the **resistor.**

(d) **Measure** and **record** the **voltage** across the **resistor**.

(e) **Calculate** the **voltage** across the **inductor**.

(f) **Measure** and **record** the **voltage** across the **inductor**.

(g) **Calculate** the **phase angle theta (θ)** for the circuit.

Inputs A and B of the oscilloscope are connected to show the $\angle\theta$ for the circuit. **Examine** the phase shift between the two waveforms displayed

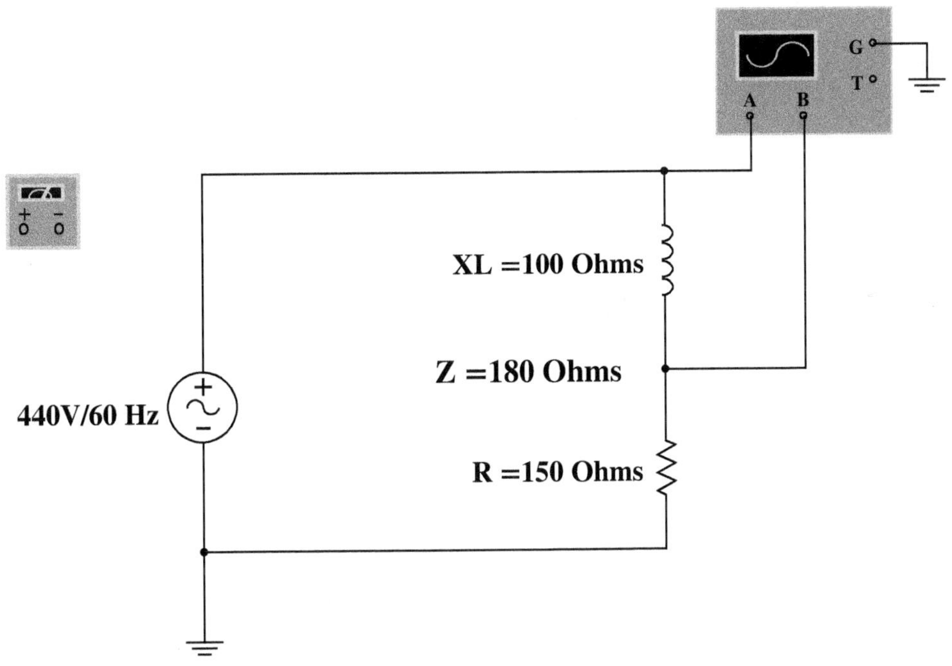

<u>**Answers**</u>

17-9 Open the Multisim circuit file CH17_09.

(a) **Record** the value of the current **(I),** applied voltage **(E_T),** voltage across the resistor **(E_R),** and voltage across the inductor **(E_L).**

(b) **Calculate** the **apparent power (VA)** of the circuit.

(c) **Calculate** the **inductive reactive power (VARs)** of the circuit.

(d) **Calculate** the **true power (W)** of the circuit.

(e) **Calculate** the **power factor (PF)** of the circuit.

(f) **Connect** the wattmeter to **measure** and **record** the **circuit power** and **power factor.**

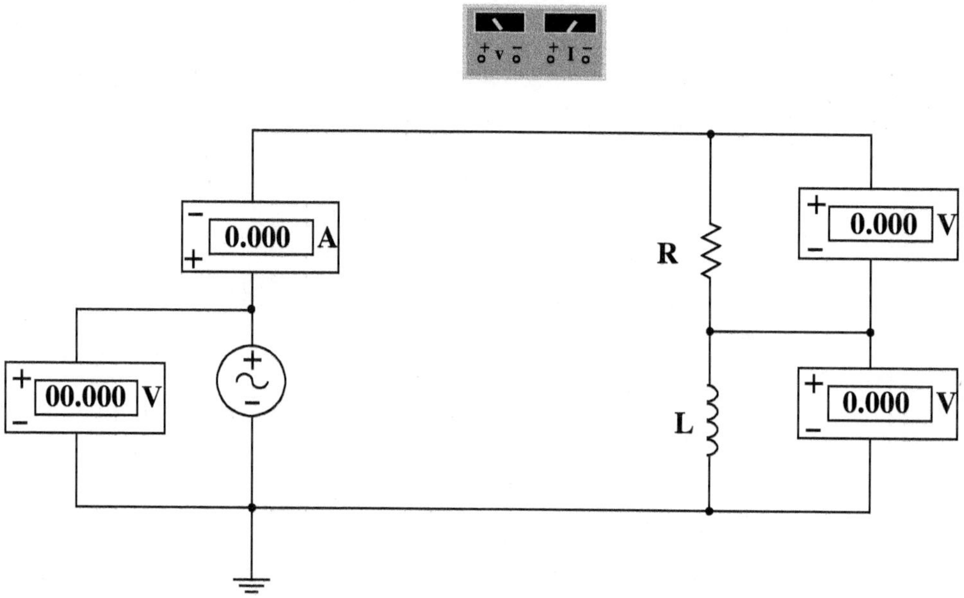

<u>Answers</u>

17-10 Open the Multisim circuit file CH17_10.

Use the multimeter, wattmeter, and whatever math calculations may be necessary to complete the table of values for the circuit shown.

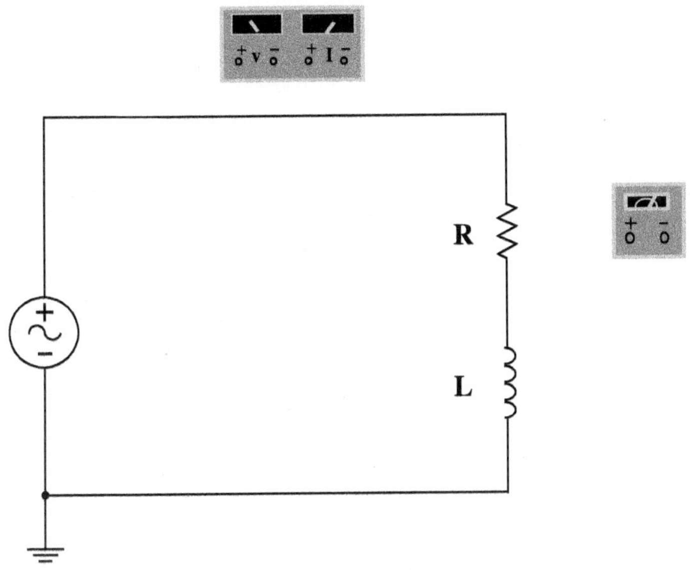

Answers

<table>
<tr><th></th><th>E</th><th>I</th><th>R-X_L-Z</th><th>W-VA-VARs</th><th>∠θ</th><th>PF</th></tr>
<tr><td>R</td><td></td><td></td><td></td><td></td><td></td><td rowspan="2">N/A</td></tr>
<tr><td>L</td><td></td><td></td><td></td><td></td><td></td></tr>
<tr><td>Total</td><td></td><td></td><td></td><td></td><td></td><td></td></tr>
</table>

17-11 Open the Multisim circuit file CH17_11.

(a) **Calculate** the **impedance (Z)** for the circuit.

(b) **Calculate** the **current (I)** for the circuit.

(c) **Measure** and **record** the value of the **current** flow.

(d) **Calculate** the **voltage** across the **resistor (E_R)** and the **capacitor (E_C).**

(e) **Measure** and **record** the **voltage** across the **resistor (E_R)** and the **capacitor (E_C).**

(f) **Calculate** the **phase angle theta (θ)** for the circuit

Inputs A and B of the oscilloscope are connected to show the ∠θ for the circuit. **Examine** the phase shift between the two waveforms displayed.

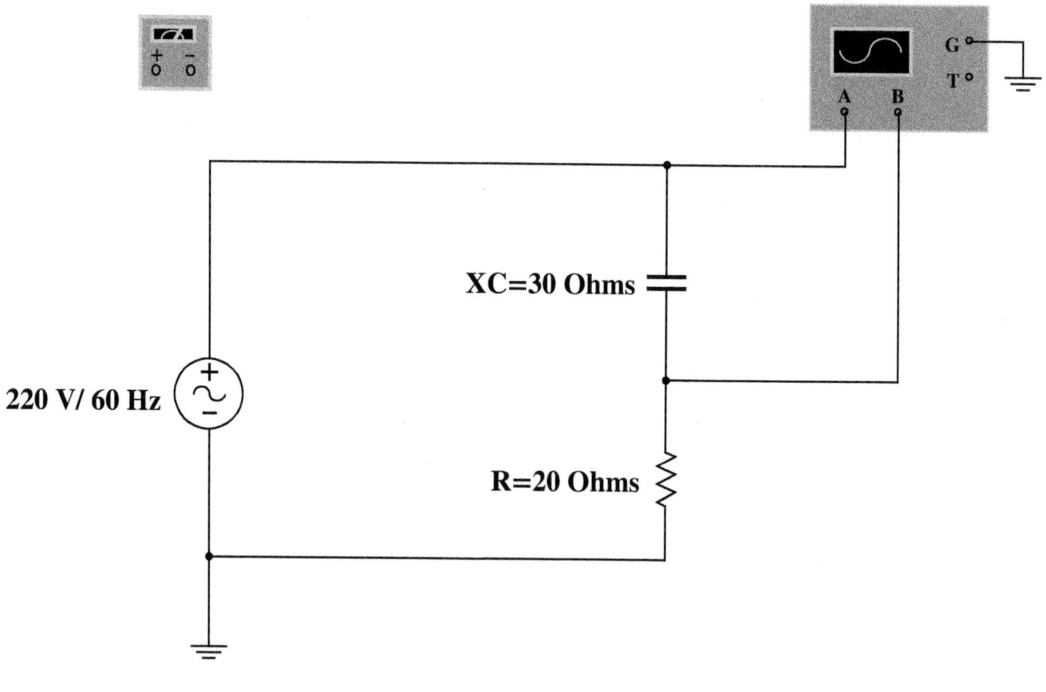

Answers

17-12 Open the Multisim circuit file CH17_12.

(a) **Calculate** the **current (I)** for the circuit.

(b) **Measure** and **record** the value of the **current (I)** flow.

(c) **Calculate** the **voltage** across the **resistor**.

(d) **Measure** and **record** the **voltage** across the **resistor**.

(e) **Calculate** the **voltage** across the **capacitor**.

(f) **Measure** and **record** the **voltage** across the **capacitor**.

(g) **Calculate** the **phase angle theta (θ)** for the circuit.

Inputs A and B of the oscilloscope are connected to show the $\angle\theta$ for the circuit. **Examine** the phase shift between the two waveforms displayed.

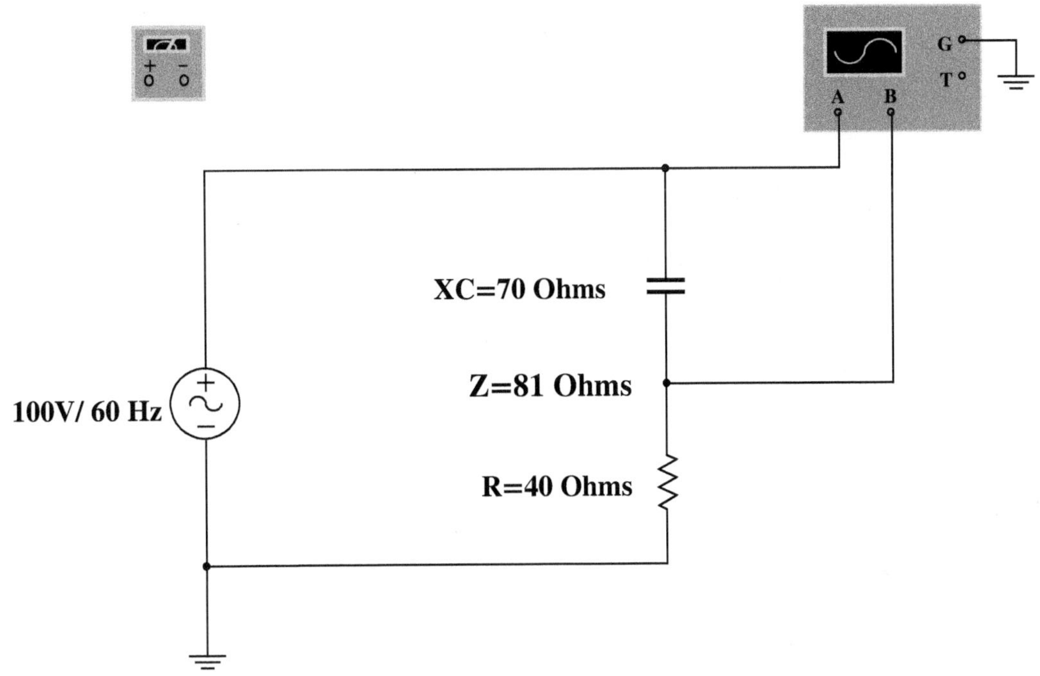

Answers

17-13 Open the Multisim circuit file CH17_13.

(a) **Record** the value of the **current (I)**, applied **voltage (E_T), voltage (E_R)** across the resistor, and **voltage (E_L)** across the inductor.

(b) **Calculate** the value of the **apparent power (VA).**

(c) **Calculate** the value of the **capacitive reactive power (VARs).**

(d) **Calculate** the value of the **true power (W).**

(e) **Calculate** the **power factor (PF)** of the circuit.

(f) **Connect** the wattmeter to **measure** and **record** the **circuit power** and **power factor.**

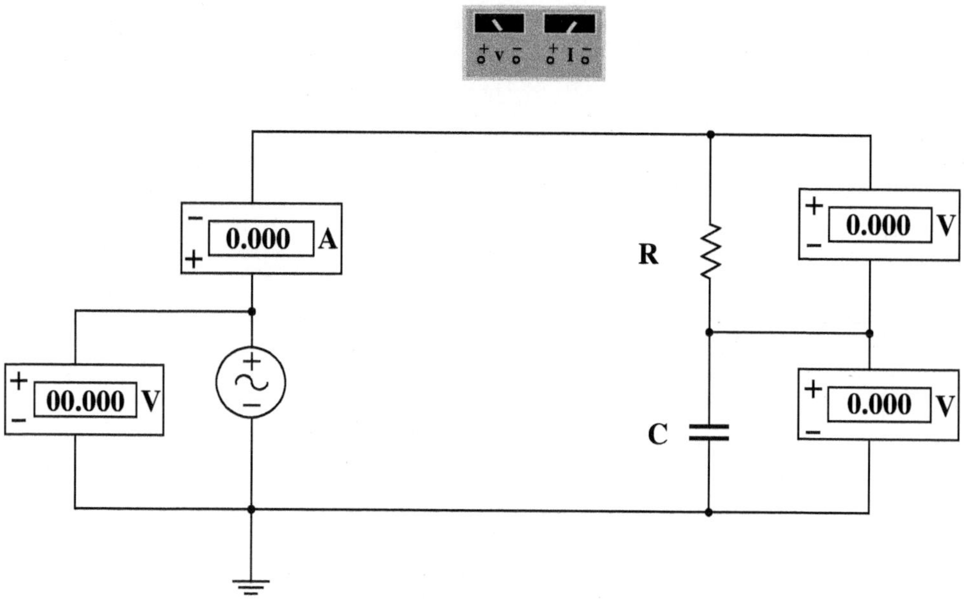

Answers

17-14 Open the Multisim circuit file CH17_14.

Use the multimeter, wattmeter, and whatever math calculations may be necessary to complete the table of values for the circuit shown.

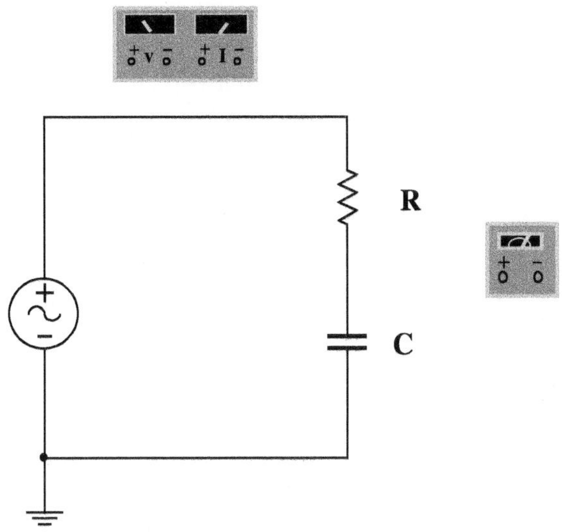

Answers

	E	I	R-X_C-Z	W-VA-VARs	$\angle\theta$	PF
R						
C						N/A
Total						

17-15 Open the Multisim circuit file CH17_15.

Inputs A and B of the oscilloscope are connected to show the **inductor and capacitor voltage waveforms**. Turn the simulation on and examine the phase relationship between the two waveforms displayed. By how many **degrees** are the two voltages **out-of-phase**?

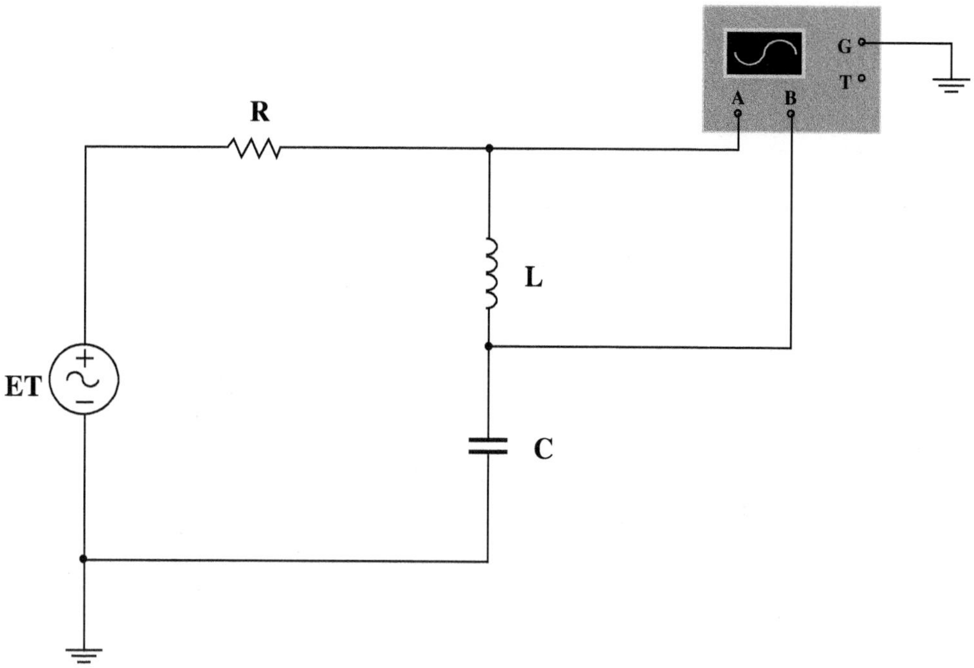

<u>Answer</u>

17-16 Open the Multisim circuit file CH17_16.

(a) **Record** the value of the **voltage** across the **resistor, inductor,** and **capacitor.**

(b) **Calculate** the value of the **applied voltage** based on the voltmeter readings.

(c) **Measure** and **record** the value of the **applied voltage**.

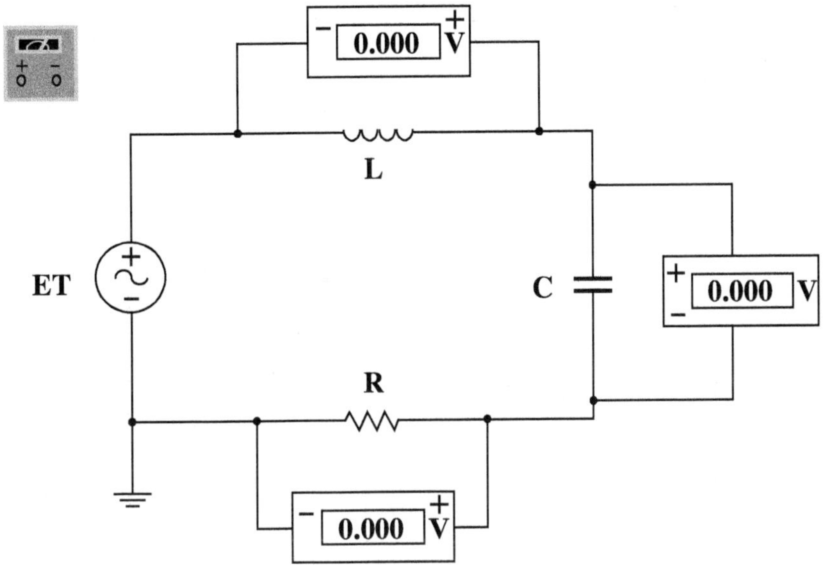

<u>**Answers**</u>

17-17 Open the Multisim circuit file CH17_17.

(a) **Calculate** the value of the **impedance (Z).**

(b) **Calculate** the value of the **current (I).**

(c) **Measure** and **record** the **current**.

(d) **Calculate** the value of the **voltage** across the **inductor, capacitor, and resistor.**

(e) **Measure** and **record** the **inductor, capacitor, and resistor voltage** values.

(f) **Calculate** the phase angle **theta (θ)** for the circuit. Inputs A and B of the oscilloscope are connected to show the $\angle\theta$ for the circuit. **Examine** the phase shift between the two waveforms displayed.

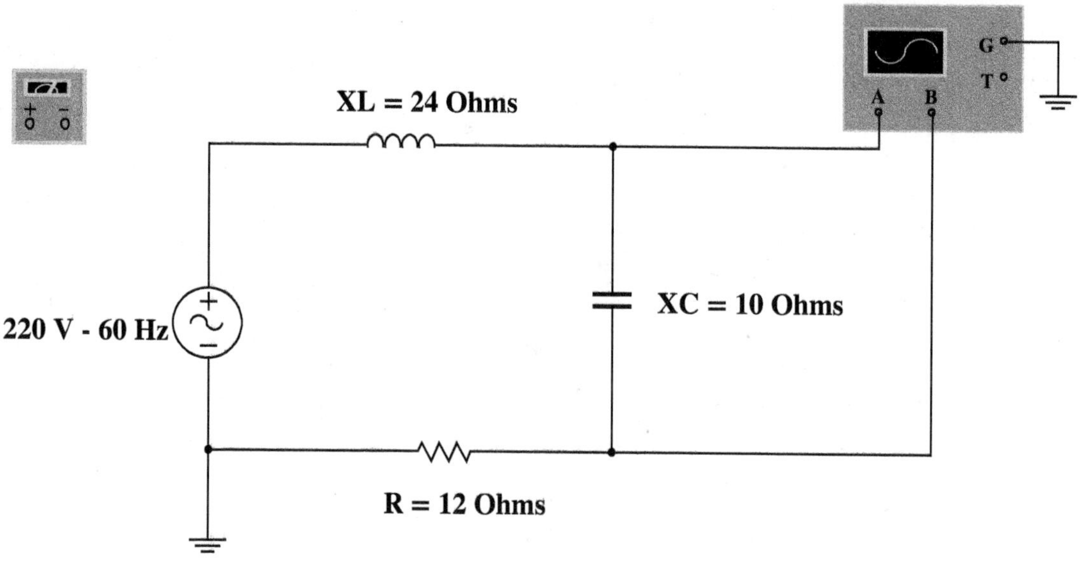

<u>Answers</u>

17-18 Open the Multisim circuit file CH17_18.

Turn the simulation on, and based on the circuit voltmeter and ammeter reading:

(a) **Calculate** the **true power (W)**

(b) **Calculate** the **apparent power (VA)**

(c) **Calculate** the **VARs inductive**

(d) **Calculate** the **VARs capacitive**

(e) **Calculate** the net **VARs**

(f) **Calculate** the **power factor (PF)**

(g) **Connect** the wattmeter to **measure** and **record** the **true power** and **power factor** of the circuit.

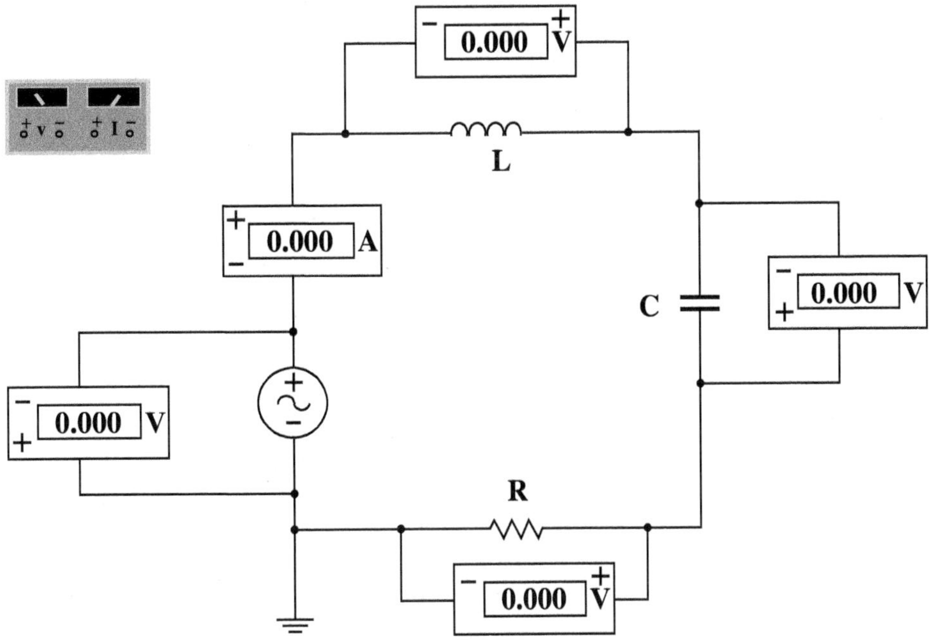

Answers

17-19 Open the Multisim circuit file CH17_19.

Turn the simulation on, and based on the circuit voltmeter and ammeter reading:

(a) **Calculate** the **apparent power (VA)**

(b) **Calculate** the **true power (W)**

(c) **Calculate** the **power factor (PF)**

(d) **Connect** the wattmeter to **measure** and **record** the **true power** and **power factor** of the circuit.

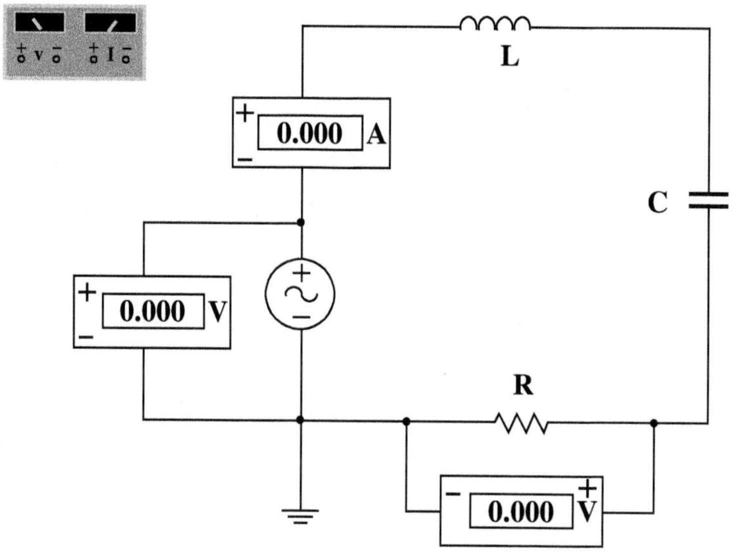

Answers

17-20 Open the Multisim circuit file CH17_20.

Use the **multimeter, wattmeter**, and whatever **math calculations** may be necessary to **complete the table** of values for the circuit shown.

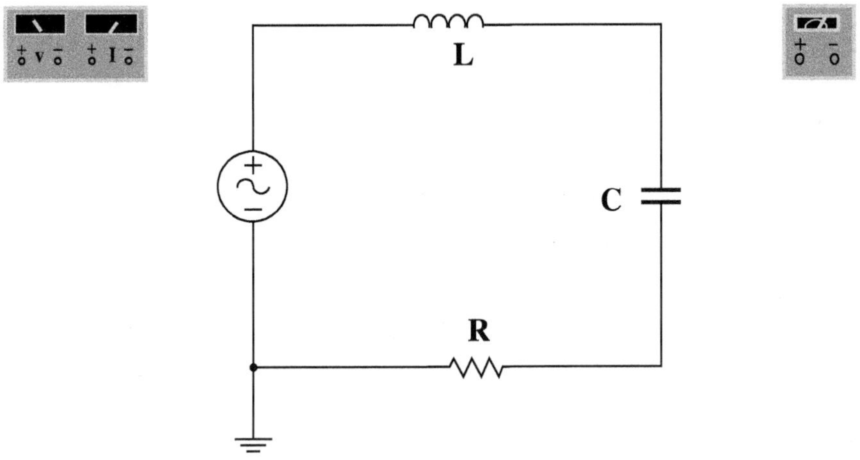

<u>**Answers**</u>

	E	**I**	**R /X_L/X_C/Z**	**W/ VA /VARs**	**PF**
R					N/A
L					
C					
Total					

17-21 Open the Multisim circuit file CH17_21.

For the **series resonant RLC circuit** shown, **measure** and **record**:

(a) the applied **voltage (E_T)**

(b) **current** flow **(I)**

(c) the **voltage** across the resistor **(E_R)**

(d) the **voltage** across the inductor **(E_L)**

(e) the **voltage** across the **capacitor (E_C)**

(f) the **voltage** across the series-connected inductor and capacitor **(E_L-to-E_C)**

(g) **true power** in watts

(h) **power factor (PF)**

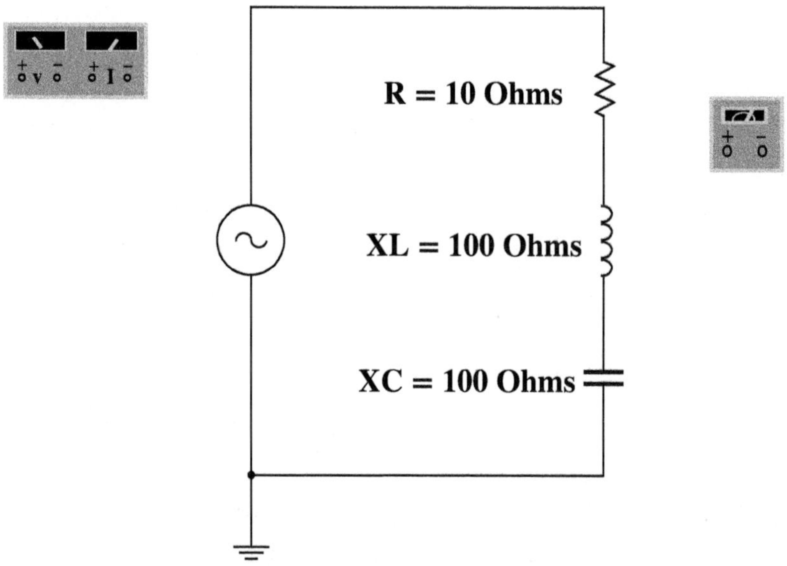

Answers

17-22 Open the Multisim circuit file CH17_22.

(a) The oscilloscope is connected to view the applied **voltage E_T** and the voltage across **resistor E_R** for the series resonant circuit shown. Turn the simulation on and **view the two waveforms**. What is the **phase relationship** between the two voltages?

(b) What is the value of the circuit **impedance Z**? (Give a reason for your answer.)

(c) **Calculate** the value of the **current** flow.

(d) **Measure** and **record** the value of the **current** flow.

(e) **Calculate** the value of the **voltage E_R** across the resistor.

(f) **Measure** and **record** the value of the **voltage E_R** across the resistor.

(g) **Calculate** the value of the **voltage E_L** across the inductor.

(h) **Measure** and **record** the value of the **voltage E_L** across the inductor.

(i) **Calculate** the value of the **voltage E_C** across the capacitor.

(j) **Measure** and **record** the value of the **voltage E_C** across the capacitor.

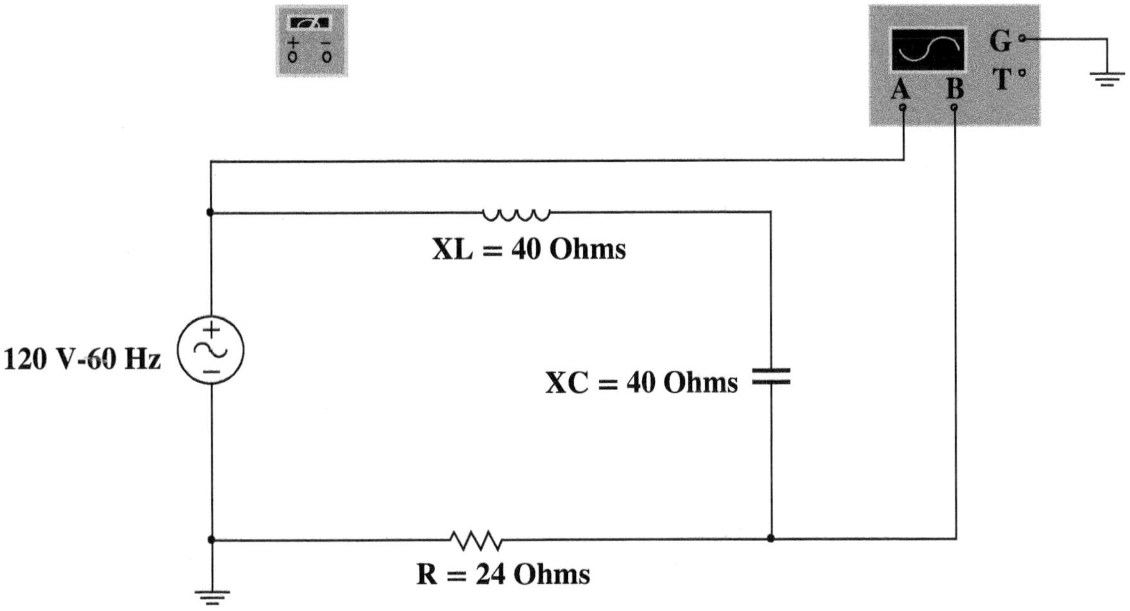

Answers

17-23 Open the Multisim circuit file CH17_23.

(a) **Calculate** the **resonant frequency** for the series RLC circuit based on the given values of L and C.

(b) When the applied 240-volts is set to the resonant frequency, calculate the value of the **current** flow and the **voltage across R, L, and C.**

(c) Double-click on the applied voltage source and **set the 240-volts to the resonant frequency. Measure** and **record** the value of the **current** flow and the **voltage across R, L, and C.**

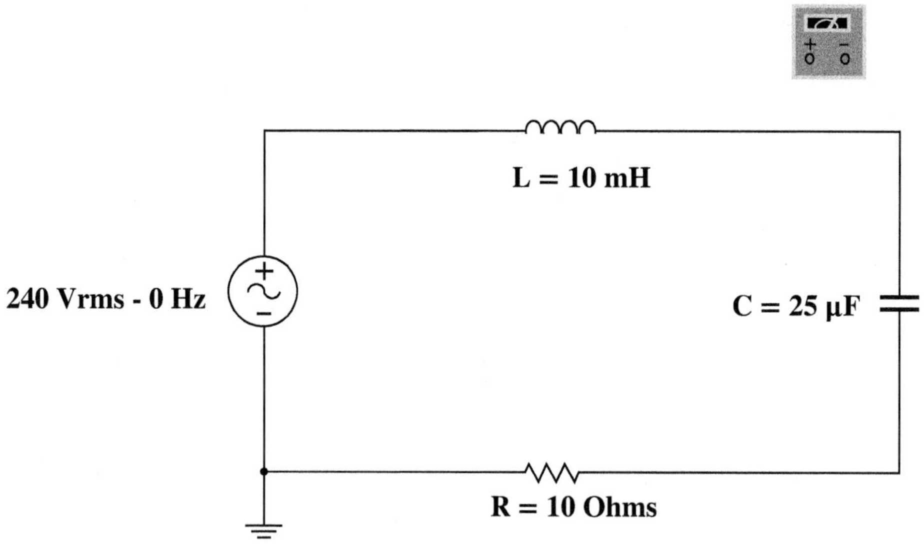

<u>Answers</u>

17-24 Open the Multisim circuit file CH17_24.

The circuit is set for its **resonant frequency of 159.15 Hz.**

(a) **Record** the value of the **current** flow and the **voltage across R, L, and C.**

(b) **Increase the frequency** of the voltage source **above resonance** to **200 Hz** and answer the following:

i. Does the original **current increase or decrease? Why?**

ii. Does the original **voltage E_R increase or decrease? Why?**

iii. Does the original **voltage E_L increase or decrease? Why?**

iv. Does the original **voltage E_C increase or decrease? Why?**

v. Is the circuit now **inductive or capacitive** in nature? **Why?**

(c) **Decrease the frequency** of the voltage source **below resonance** to **100 Hz** and answer the following:

i. Does the original **current increase or decrease? Why?**

ii. Does the original **voltage E_R increase or decrease? Why?**

iii. Does the original **voltage E_L increase or decrease? Why?**

iv. Does the original **voltage E_C increase or decrease? Why?**

v. Is the circuit now **inductive or capacitive** in nature? **Why?**

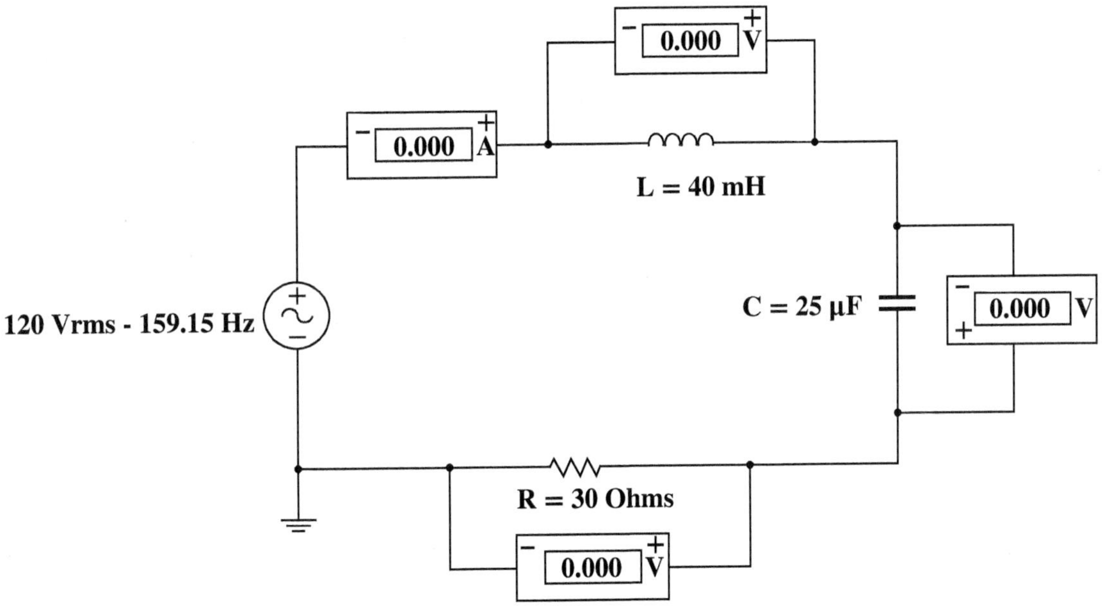

<u>Answers</u>

Chapter 18 Resistive, Inductive, Capacitive (RLC) Parallel Circuits

18-1 Open the Multisim circuit file CH18_01.

(a) **Measure** and **record** the value of $\mathbf{E_T}$, $\mathbf{E_R}$, **and** $\mathbf{E_L}$.

(b) **Measure** and **record** the value of $\mathbf{I_T}$, $\mathbf{I_R}$, **and** $\mathbf{I_L}$.

(c) Based on your readings, **calculate** the **impedance (Z)** of the circuit.

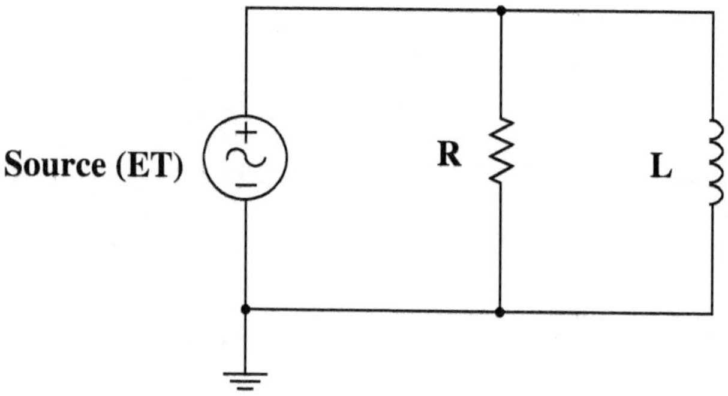

Answers

18-2 Open the Multisim circuit file CH18_02.

(a) **Calculate** the value of $\mathbf{I_R}$.

(b) **Measure** and **record** the value of $\mathbf{I_R}$.

(c) **Calculate** the value of $\mathbf{I_L}$.

(d) **Measure** and **record** the value of $\mathbf{I_L}$.

(e) **Calculate** the value of $\mathbf{I_T}$.

(f) **Measure** and **record** the value of $\mathbf{I_T}$.

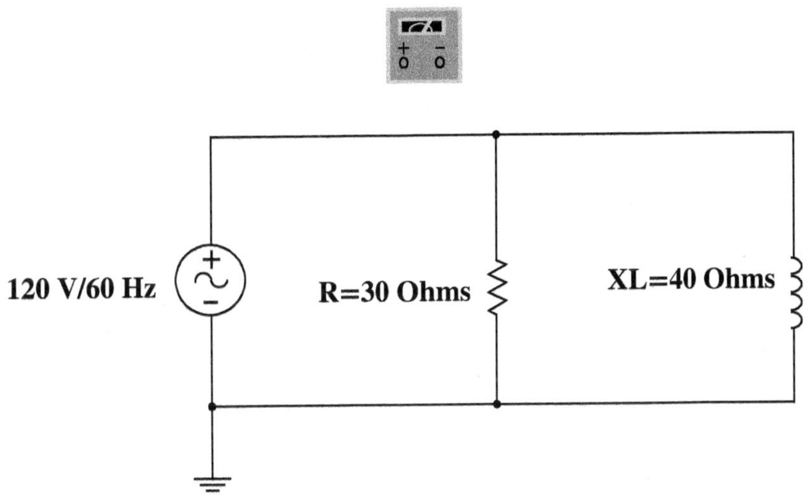

<u>Answers</u>

18-3 Open the Multisim circuit file CH18_03.

(a) **Calculate** the **impedance (Z)** based on the given R and X_L values.

(b) **Calculate** the **total current (I_T)** based on the calculated impedance value.

(c) **Measure** and **record** the value of I_T.

(d) **Calculate** the **current (I_R)** through the resistor.

(e) **Measure** and **record** the value of I_R.

(f) **Calculate** the **current (I_L)** through the inductor.

(g) **Measure** and **record** the value of I_L.

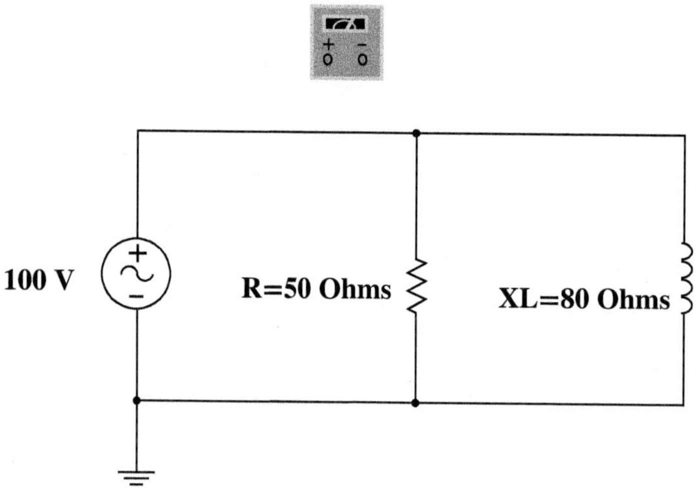

<u>Answers</u>

18-4 Open the Multisim circuit file CH18_04.

(a) **Record** the value of the **applied voltage (E)**, **total current (I_T)**, **current** through the resistor **(I_R)**, and the **current** through the inductor **(I_L)**.

(b) **Calculate** the value of the **apparent power (VA).**

(c) **Calculate** the value of the **inductive reactive power (VARs).**

(d) **Calculate** the value of the **true power (W).**

(e) **Measure** and **record** the **total wattage** of the circuit.

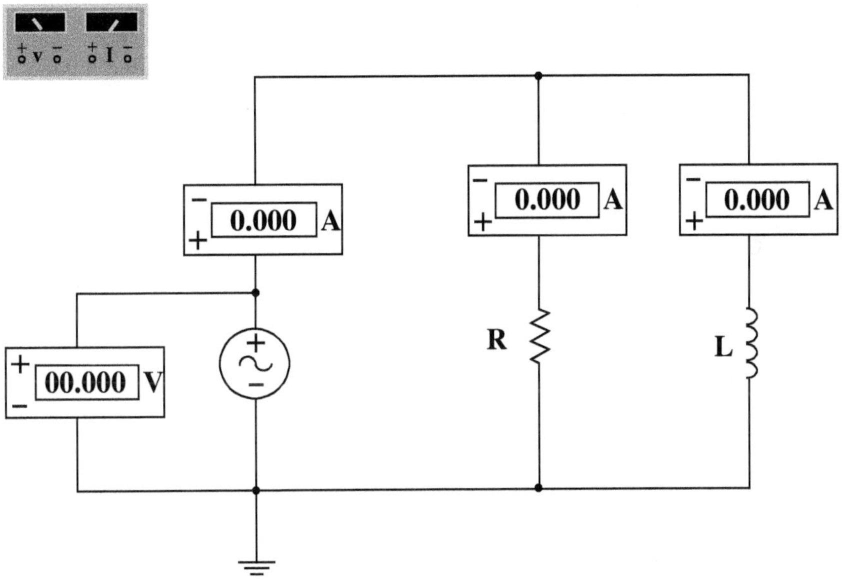

Answers

18-5 Open the Multisim circuit file CH18_05.

Use the **multimeter** and whatever other **math calculations** may be necessary to **complete the table** of values for the circuit shown.

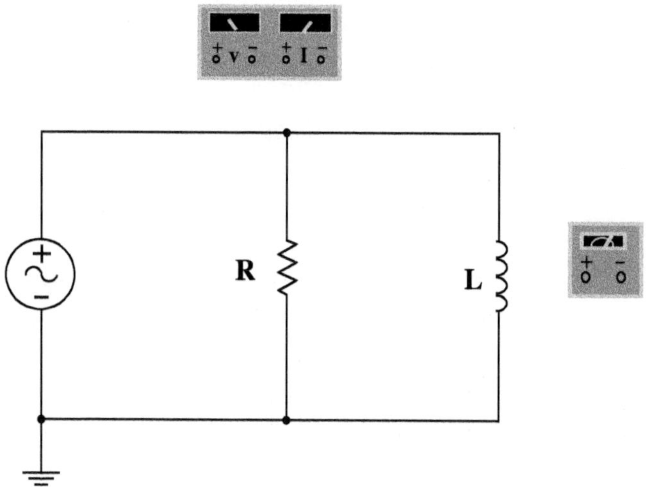

Answers

	E	I	R /X_L/Z	W/VA/VARs	∠θ	PF
R						
L						N/A
Total						

18-6 Open the Multisim circuit file CH18_06.

Complete the table for all **given and unknown quantities** for the parallel RL circuit shown. Use the **multimeter** to check as many of **your calculations** as possible.

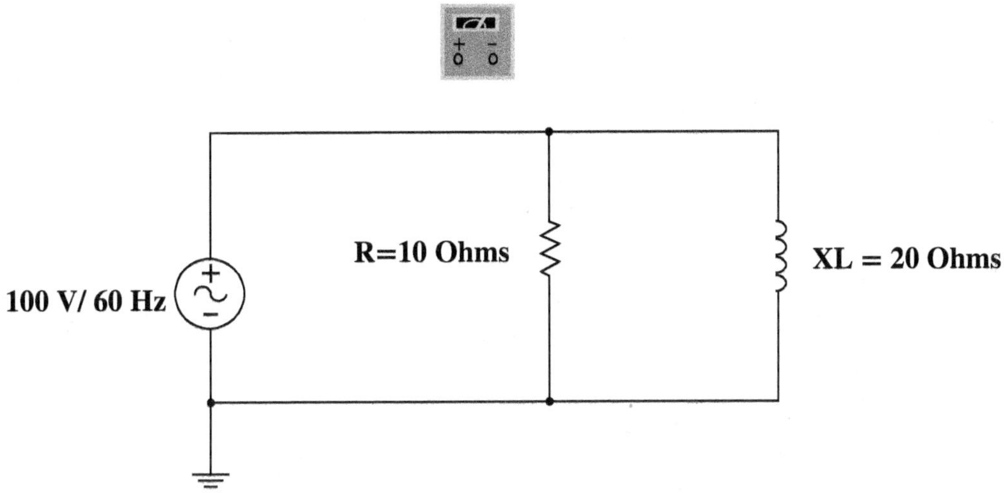

Answers

	E	I	$R/X_L/Z$	W/VA/VARs
R				
L				
Total				

18-7 Open the Multisim circuit file CH18_07.

(a) **Calculate** current flow I_R.

(b) **Measure** and **record** current flow I_R.

(c) **Calculate** current flow I_C.

(d) **Measure** and **record** current flow I_C.

(e) **Calculate** current flow I_T.

(f) **Measure** and **record** current flow I_T.

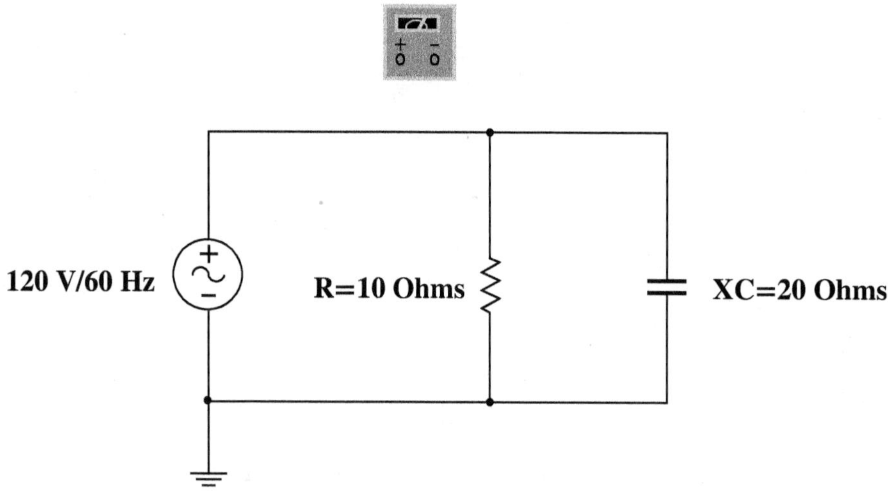

Answers

18-8 Open the Multisim circuit file CH18_08.

(a) **Record** the value of the applied **voltage (E)**, total **current (I_T)**, current through the **resistor (I_R)**, and the current through the **capacitor (I_C).**

(b) **Calculate** the value of the **apparent power (VA).**

(c) **Calculate** the value of the **inductive reactive power (VARs).**

(d) **Calculate** the value of the **true power (W).**

(e) **Calculate** the value of the **power factor (PF).**

(f) **Connect the wattmeter** into the circuit and record the **circuit wattage** and **power factor readings**.

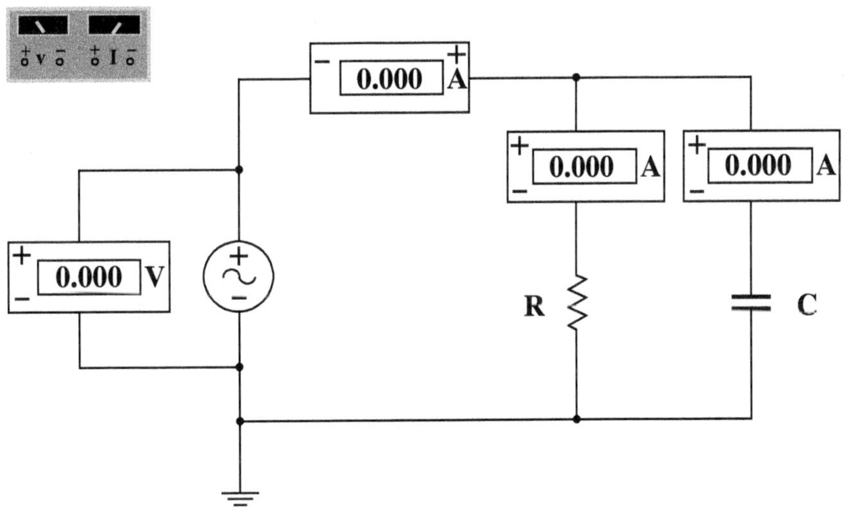

<u>Answers</u>

18-9 Open the Multisim circuit file CH18_09.

Complete the table for all given and unknown quantities for the parallel RC circuit shown. Use the **multimeter and wattmeter** to check as many of **your calculations** as possible.

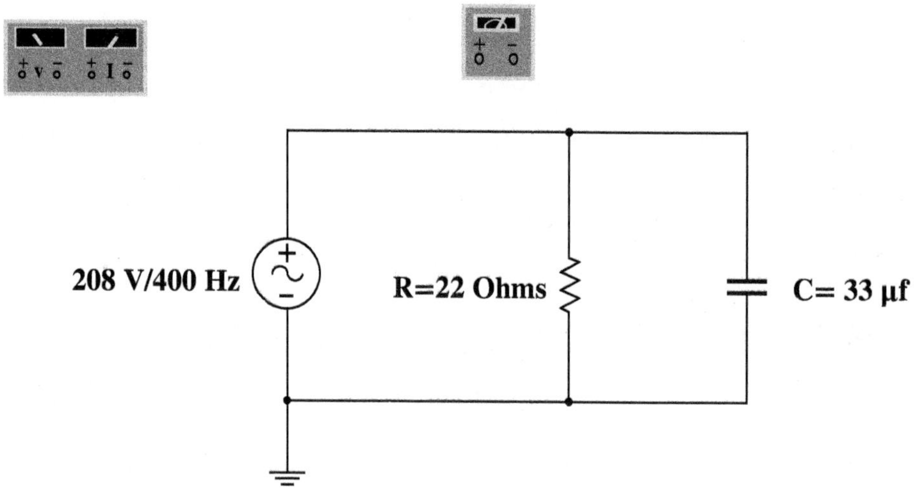

<u>Answers</u>

	E	I	C	R /X_C/Z	W/VA/VARs	∠θ	PF
R			N/A				
C							NA
Total			N/A				

18-10 Open the Multisim circuit file CH18_10.

(a) **Record** the **wattage** and **power factor readings.**

(b) **Calculate** the **current** flow through the **resistor**. Verify your answer by **measuring** the resistor current.

(c) **Calculate** the **resistance** value of the **resistor**. Verify your answer by **measuring** the resistance value.

(d) **Calculate** the **capacitive reactance (X_C)** of the capacitor.

(e) **Calculate** the **current** flow through the **capacitor**. Verify your answer by measuring the capacitor current.

(f) **Calculate** the value of the **line current**. Verify your answer by measuring the line current.

(g) **Calculate** the value of the **apparent power (VA).**

(h) **Calculate** the % **PF.**

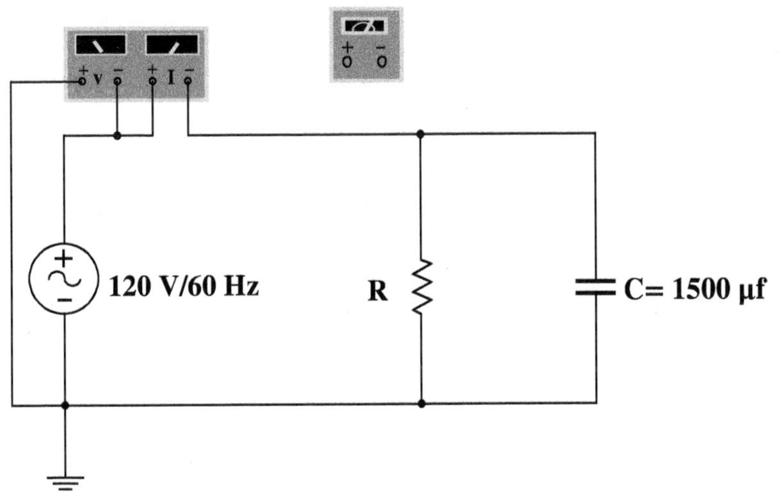

<u>Answers</u>

18-11 Open the Multisim circuit file CH18_11.

Complete the table for all **given and unknown quantities** for the parallel RC circuit shown. Use the multimeter and wattmeter to **verify** as many of your calculations as possible.

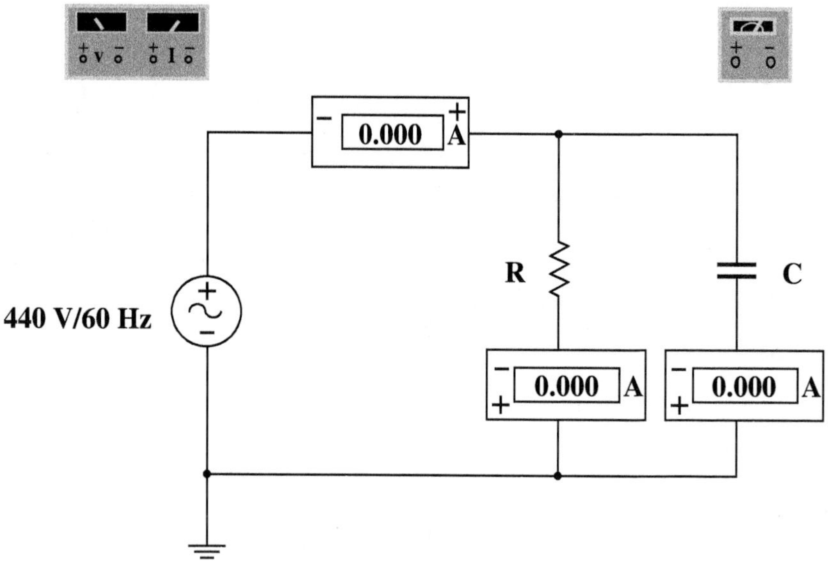

<u>Answers</u>

	E	I	C	R /X_C/Z	W/VA/VARs	∠θ	PF
R			N/A				
C							N/A
Total			N/A				

18-12 Open the Multisim circuit file CH18_12.

(a) **Record** the value of the **current** flow through the **inductor** and **capacitor.**

(b) **Calculate** the value of the **line current**.

(c) **Measure** and **record** the value of the **line current**.

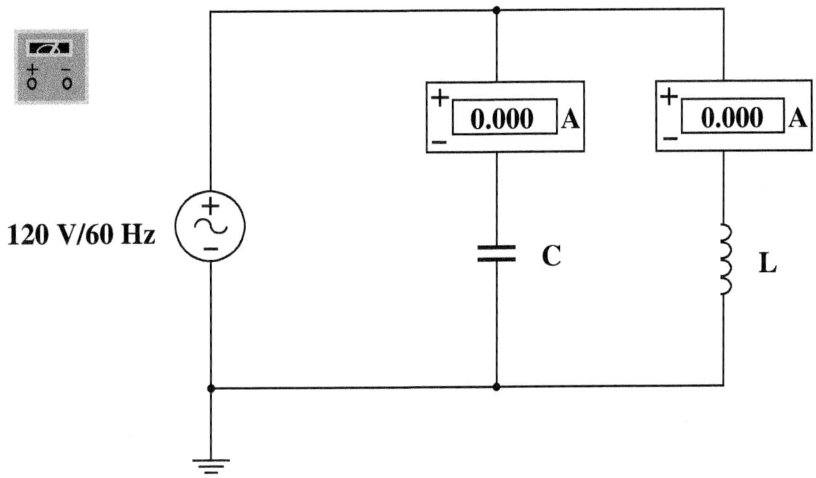

<u>**Answers**</u>

18-13 Open the Multisim circuit file CH18_13.

(a) **Calculate** the **impedance (Z)** of the circuit based on the reactance values $\mathbf{X_C}$ and $\mathbf{X_L}$.

(b) **Calculate** the **line current** flow based on your impedance calculation.

(c) **Measure** and **record** the value of the line current.

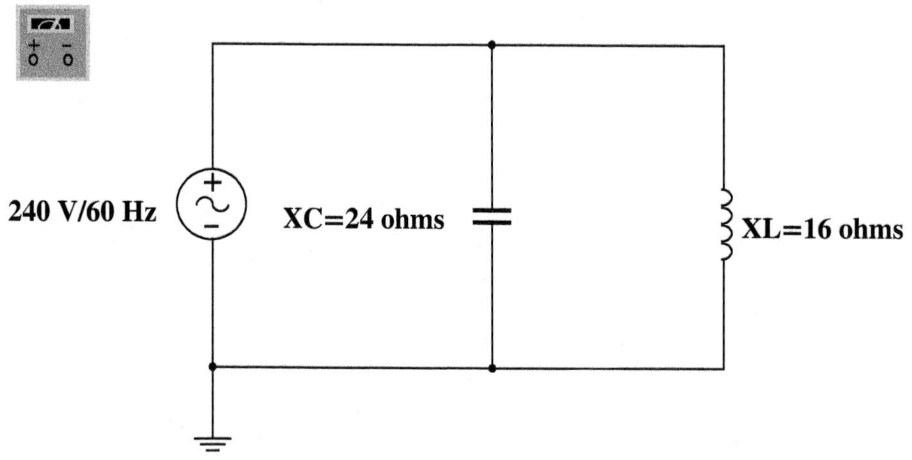

<u>**Answers**</u>

18-14 Open the Multisim circuit file CH18_14.

(a) **Record** the value of the **current** flow through the **inductor**, **capacitor,** and **line.**

(b) **Determine** the **apparent power (VA).**

(c) **Determine** the **reactive power VARs).**

(d) **Determine** the **true power (W).**

(e) **Determine** the **power factor (PF).**

(f) **Measure** and **record** the circuit **watts** and **power factor**.

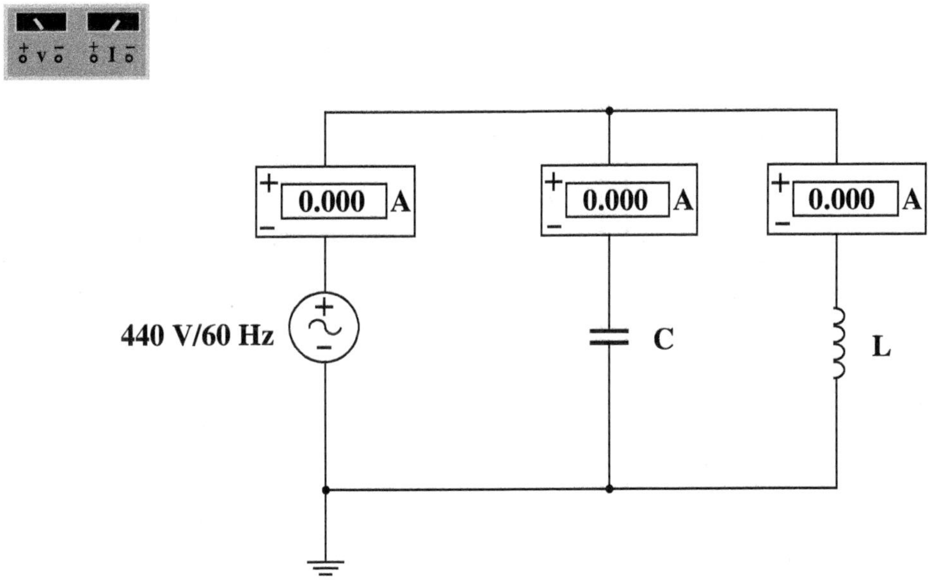

<u>Answers</u>

18-15 Open the Multisim circuit file CH18_15.

Measure and **record** the value of:

(a) The **applied voltage** and the voltage across **R, L**, and **C**. How do they **compare** in value? **Why**?

(b) The **current** flow through the **resistor**.

(c) The **current** flow through the **inductor**.

(d) The **current** flow through the **capacitor**.

(e) The **reactive current (I_X)** between **points A and B.**

(f) The total **line current**.

(g) Is the circuit **inductive or capacitive** in nature? **Why**?

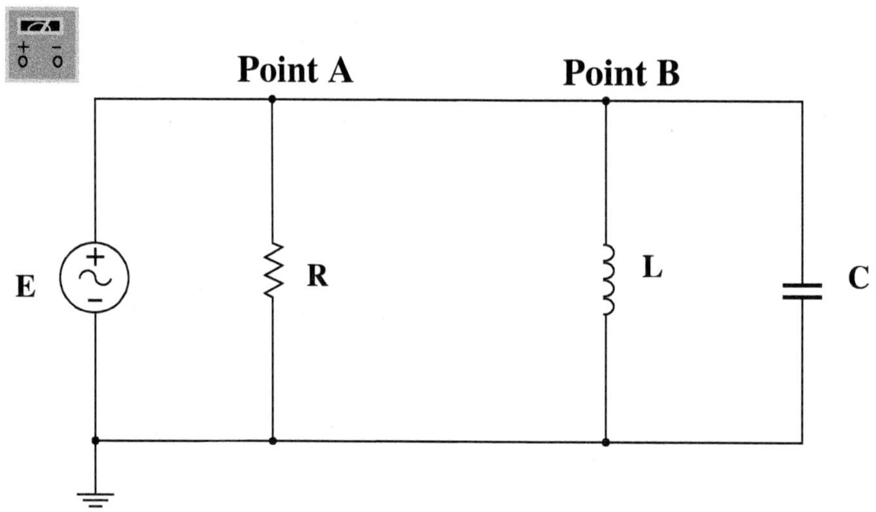

__Answers__

18-16 Open the Multisim circuit file CH18_16.

(a) **Record** the value of the **current** through the **resistor, inductor**, and **capacitor**.

(b) **Calculate** the value of the current flow between **points A and B** of the circuit.

(c) **Calculate** the value of the **line current**.

(d) **Measure** and **record** the value of the **current** between points **A and B** of the circuit.

(e) **Measure** and **record** the value of the **line current**.

(f) Is the circuit **inductive or capacitive** in nature? **Why?**

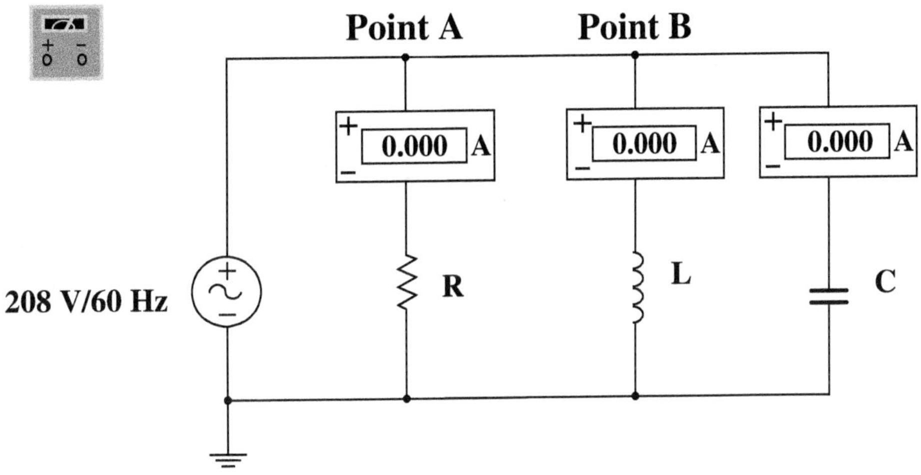

Answers

18-17 Open the Multisim circuit file CH18_17.

(a) **Calculate** the value of the **current** through the **resistor. Verify** your answer by **measuring** the resistor current.

(b) **Calculate** the value of the **current** through the **inductor. Verify** your answer by **measuring** the **inductor current.**

(c) **Calculate** the value of the **current** through the **capacitor. Verify** your answer by **measuring** the **capacitor current.**

(d) **Calculate** the value of the **net reactive current. Verify** your answer by **measuring** the **net reactive current.**

(e) **Calculate** the value of the **line current. Verify** your answer by **measuring** the **line current.**

(f) Calculate the value of the circuit **impedance (Z).**

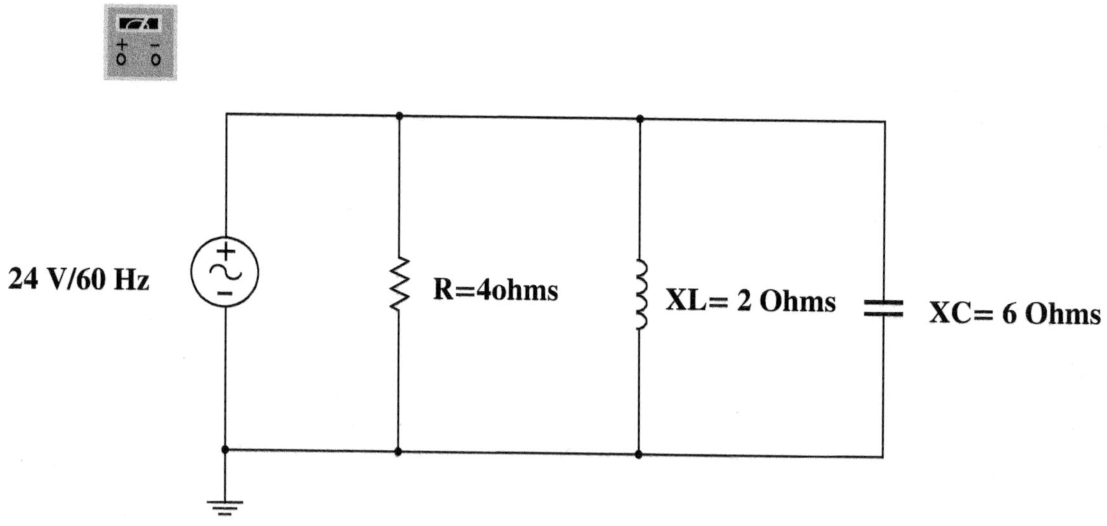

Answers

18-18 Open the Multisim circuit file CH18_18.

(a) **Record** the value of $\mathbf{I_T}$, $\mathbf{I_R}$, $\mathbf{I_X}$, $\mathbf{I_L}$, and $\mathbf{I_C}$.

(b) **Calculate** the value of the **true power (W).**

(c) **Calculate** the value of the **reactive power (VARs).**

(d) **Calculate** the value of the **apparent power (VA).**

(e) **Calculate** the **power factor (PF).**

(f) **Measure** and **record** the circuit **watts** and **power factor (PF).**

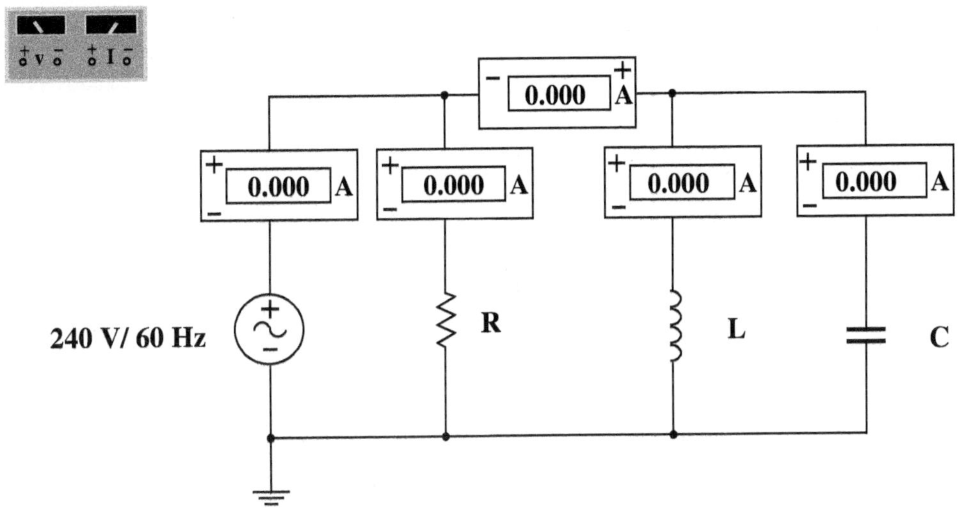

<u>Answers</u>

18-19 Open the Multisim circuit file CH18_19.

(a) **Calculate** the value of the **current** flow through the **resistor. Verify** your answer by **measuring** the resistor current.

(b) **Calculate** the value of the **current** flow through the **inductor. Verify** your answer by **measuring** the inductor current.

(c) **Calculate** the value of the **current** flow through the **capacitor**. **Verify** your answer by **measuring** the capacitor current.

(d) **Calculate** the value of the **net reactive current** flow. **Verify** your answer by **measuring** the net reactive current.

(e) **Calculate** the value of the **line current** flow. **Verify** your answer by **measuring** the line current.

(f) **Calculate** the value of the circuit **impedance (Z).**

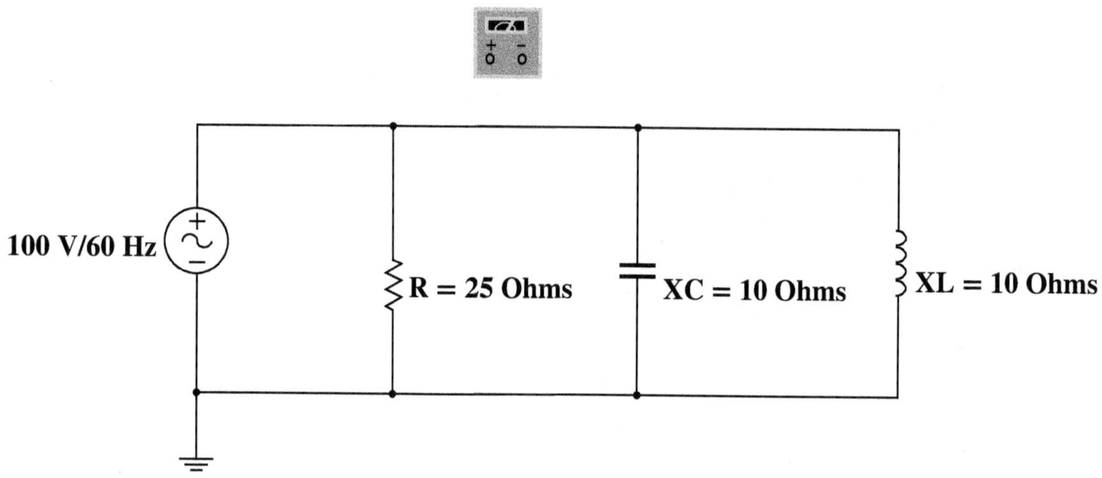

<u>**Answers**</u>

18-20 Open the Multisim circuit file CH18_20.

(a) **Record** the value of the **true power (W)** and **power factor (PF)** reading of the wattmeter.

(b) **Calculate** the value of the **line current**.

(c) **Measure** and **record** the value of the line current.

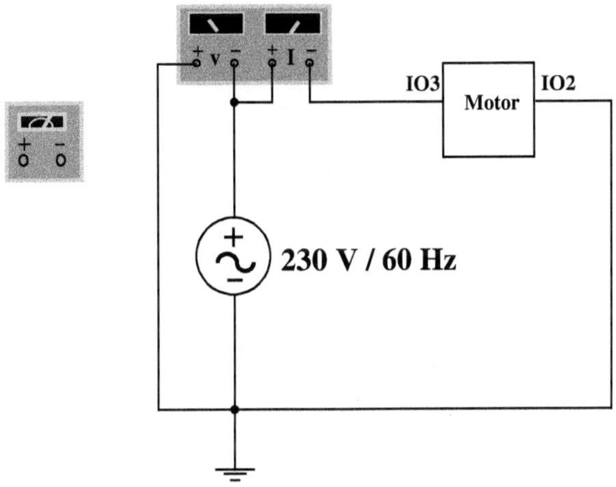

Answers

18-21 Open the Multisim circuit file CH18_21.

(a) **Record** the value of the applied **voltage** and line **current** flow.

(b) **Calculate** the **kVA** apparent power of the circuit.

(c) **Calculate** the **power factor (PF)** of the circuit.

(d) **Measure** and **record** the **wattage** and **power factor (PF)**.

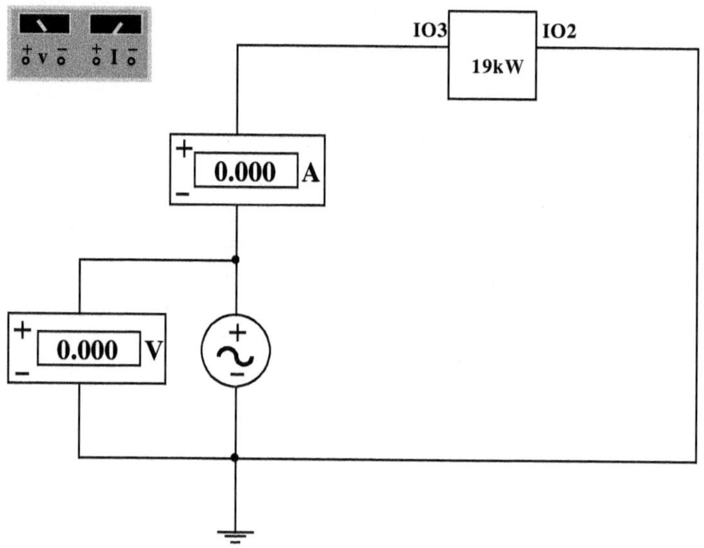

Answers

18-22 Open the Multisim circuit file CH18_22.

(a) **Connect** the **multimeter** and **wattmeter** into the circuit and **record** the value of the following:

i. **line current** flow
ii. **motor current** flow
iii. **capacitor current** flow
iv. **true power**
v. **power factor**

(b) **Repeat part (a)** with the **capacitor disconnected** from the circuit.

(c) Which **readings changed**? Which readings **stayed the same**?

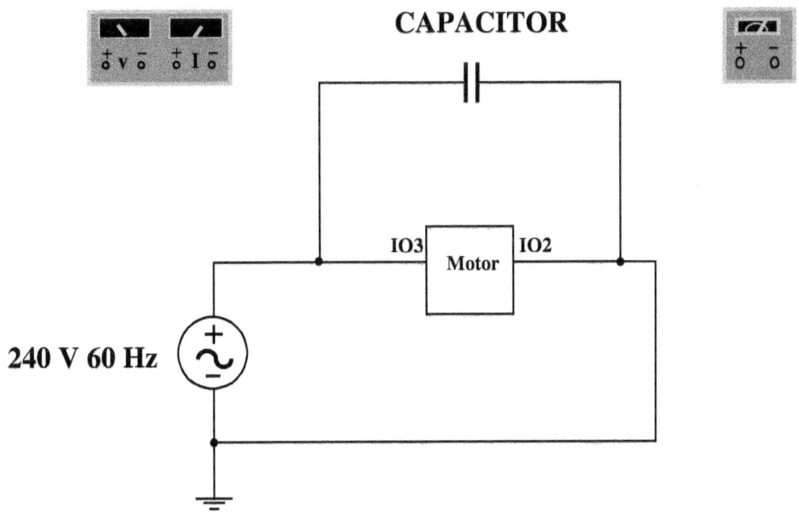

<u>**Answers**</u>

18-23 Open the Multisim circuit file CH18_23.

(a) **Record** the **wattmeter** reading, **power factor** reading, ammeter **current** reading, and voltmeter **voltage** reading for the simulated **power factor correction** circuit.

(b) **Calculate** the **apparent (VA)** and **reactive (VARS)** power.

(c) **Calculate** the **apparent (VA)** and **reactive (VARS)** power with a **capacitor** connected to achieve a **corrected power factor of 90%.**

(d) **Calculate** the **capacitive VARs** required in order to **raise** the power factor to **90%.**

(e) **Calculate** the **capacitive reactance (X_c)** of the capacitor required to **raise** the power factor to **90%.**

(f) **Calculate** the **size of capacitor (μF)**, connected in parallel with the motor, required to **raise the power factor to 90%.**

(g) **Calculate** the amount of **reduction in line current** with the power factor corrected to 90%.

(h) **Set the capacitor value** to that required to correct the power factor to 90%. Wire the capacitor in parallel with the motor and **record** the value of the **wattmeter** reading, **power factor** reading, **ammeter** reading, and **voltmeter** reading.

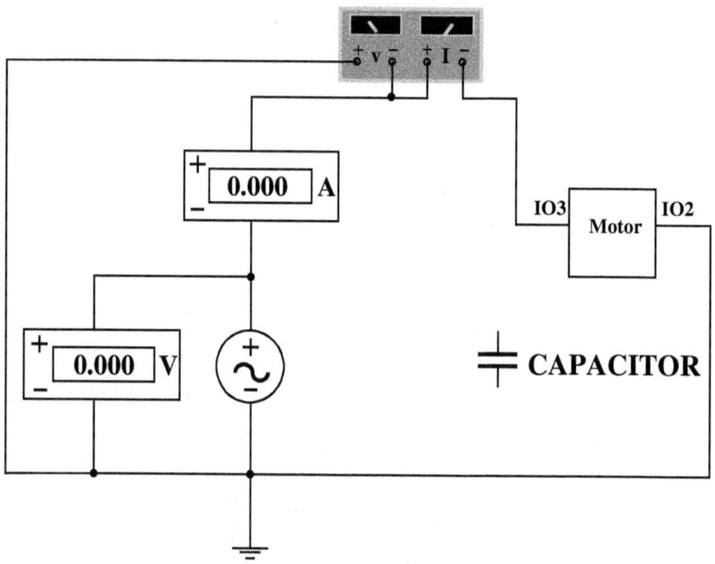

Answers

Chapter **19 Transformers**

19-1 Open the Multisim circuit file CH19_01.

(a) **Determine** the **turns ratio** of the transformer.

(b) **Calculate** the value of the **secondary voltage**.

(c) **Measure** and **record** the **secondary voltage**.

(d) **What type** of transformer would this be classified as? **Why**?

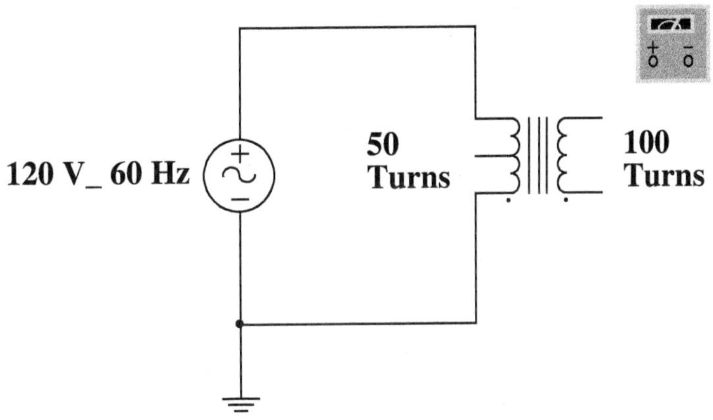

Answers

19-2 Open the Multisim circuit file CH19_02.

(a) **Determine** the **turns ratio** of the transformer.

(b) **Calculate** the value of the **secondary voltage.**

(c) **Measure** and **record** the **secondary voltage.**

(d) **What type** of transformer would this be classified as? **Why?**

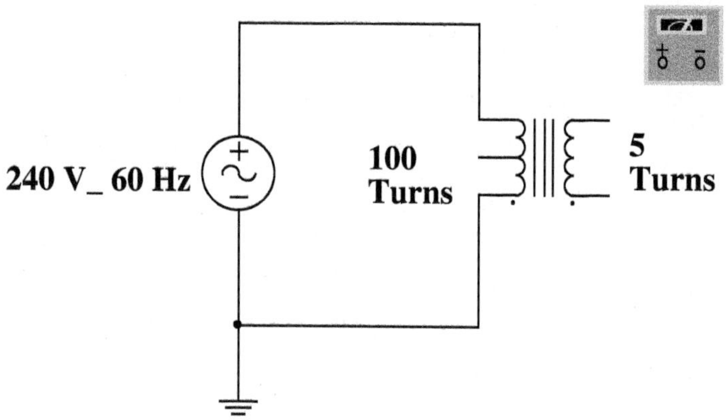

Answers

19-3 Open the Multisim circuit file CH19_03.

(a) **Record** the value of the **primary** and **secondary voltages.**

(b) **Calculate** the value of the **secondary current**.

(c) **Measure** the value of the **secondary current.**

(d) **Calculate** the **volt-amps** of the **secondary** circuit.

(e) **Calculate** the **volt-amps** of the **primary** circuit.

(f) **Calculate** the value of the **primary current.**

(g) **Measure** the value of the **primary current.**

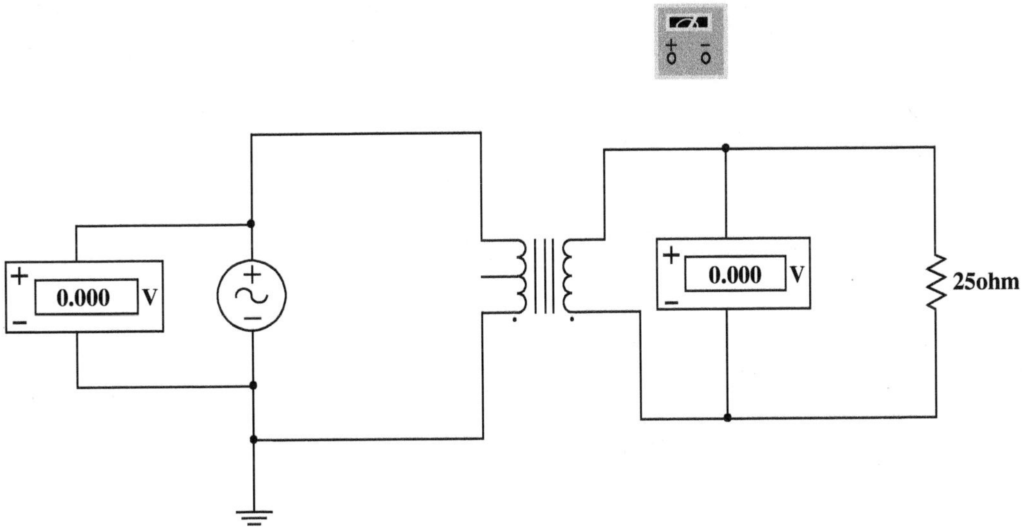

Answers

19-4 Open the Multisim circuit file CH19_04.

(a) **Record** the value of $\mathbf{E_P}$, $\mathbf{E_S}$, and $\mathbf{I_S}$.

(b) **Calculate** the value of the **primary current.**

(c) **Measure** the value of the **primary current.**

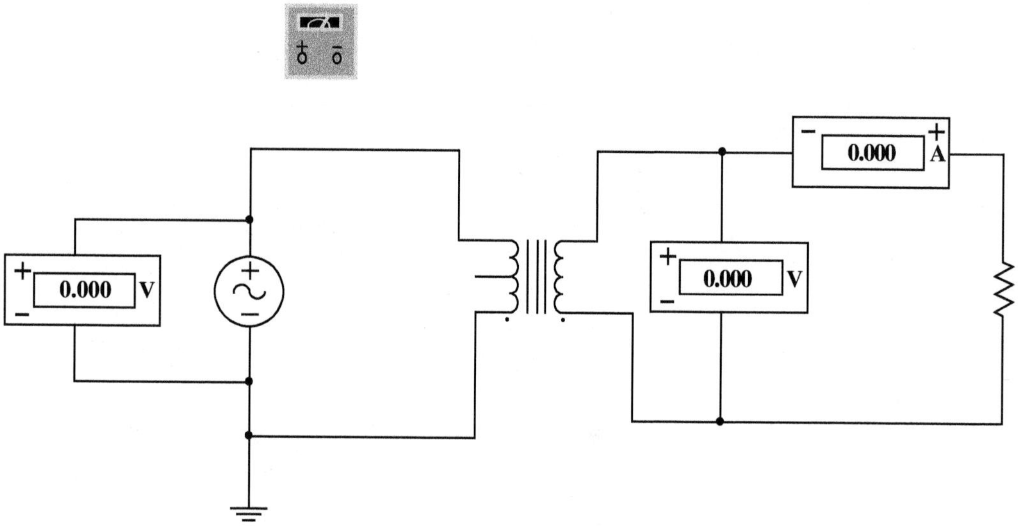

Answers

19-5 Open the Multisim circuit file CH19_05.

This circuit simulates a **polarity test** being made on a transformer.

(a) **Record** the reading of **voltmeter VM#1 and voltmeter VM#2**.

(b) **What type** of polarity (**additive or subtractive**) is indicated?

(c) What is the value of the **voltage across the secondary**? Verify this voltage value by measurement.

(d) What would the **polarity marking (X1 or X2)** for the transformer secondary terminal **adjacent to terminal H1**?

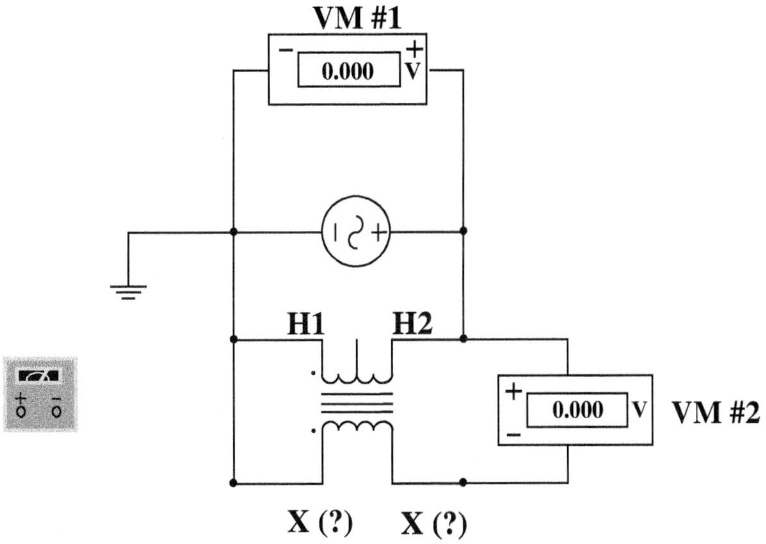

Answers

19-6 Open the Multisim circuit file CH19_06.

(a) **Complete the circuit diagram** so that 120-volts and 240-volts are supplied to their respective loads by the distribution transformer

(b) **Connect the Multisim circuit** according to the diagram.

(c) **Measure** and **record** the value of the **voltage** across Load 1, Load 2, and Load 3.

(d) **Measure** and **record** the value of the **current** through Load 1, Load 2, and Load 3.

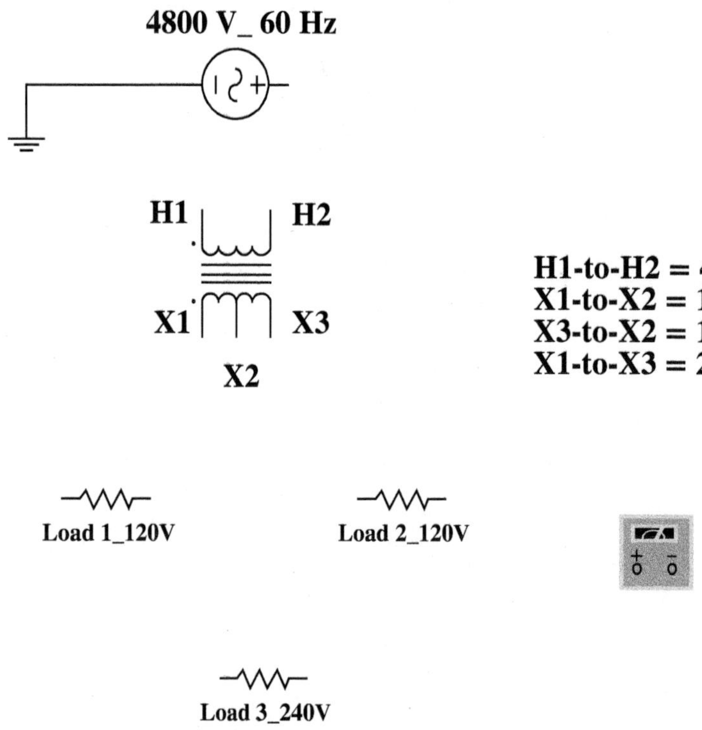

H1-to-H2 = 4800 V
X1-to-X2 = 120 V
X3-to-X2 = 120 V
X1-to-X3 = 240 V

<u>Answers</u>

19-7 Open the Multisim circuit file CH19_07.

(a) **Identify** the type of 3-phase transformer **connection** shown.

(b) **Measure** and **record** the value of the **primary line voltage.**

(c) **Measure** and **record** the value of the **primary phase voltage.**

(d) **Measure** and **record** the value of the **secondary phase voltage.**

(e) **Measure** and **record** the value of the **secondary line voltage.**

(f) **Calculate** the **turns ratio** of each transformer.

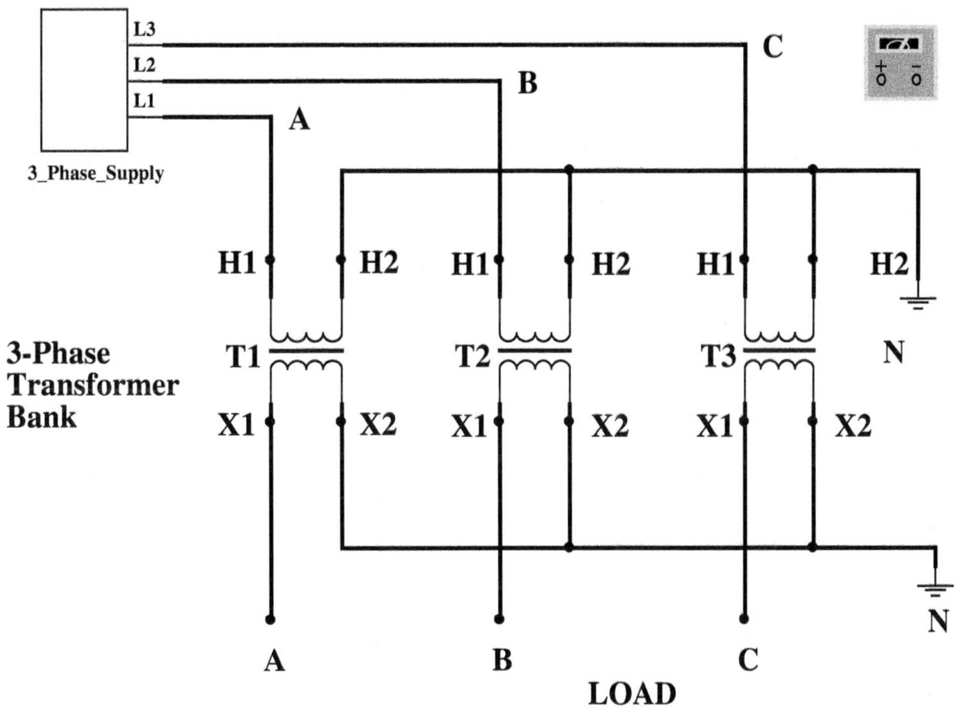

Answers

19-8 Open the Multisim circuit file CH19_08.

For the wye-wye transformer connection:

(a) **Record** the value of the **primary line voltage** as displayed on the voltmeter.

(b) **Calculate** the value of the **primary phase voltage**.

(c) **Measure** and **record** the **primary phase voltage.**

(d) **Calculate** the value of the **secondary phase voltage.**

(e) **Measure** and **record** the **secondary phase voltage.**

(f) **Calculate** the value of the **secondary line voltage.**

(g) **Measure** and **record** the **secondary line voltage.**

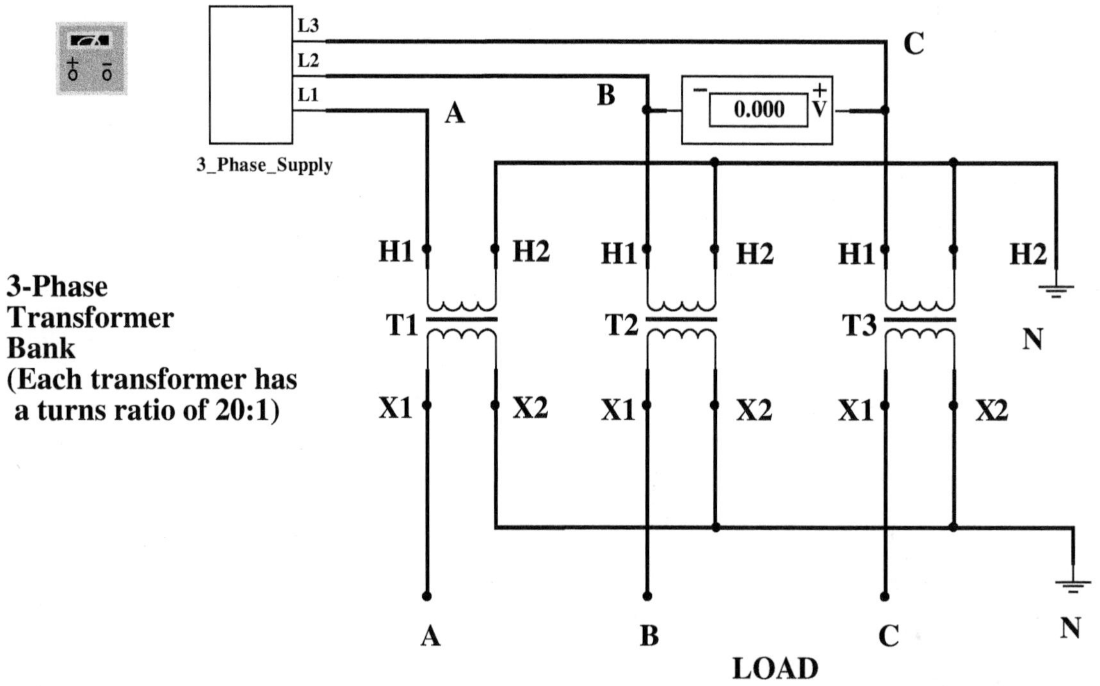

Answers

19-9 Open the Multisim circuit file CH19_09.

For the wye-wye transformer connection:

(a) **Record** the value of the **primary line-to-voltage.**

(b) **Measure** and **record** the **secondary phase voltage.**

(c) **Measure** and **record** the **secondary line voltage.**

(d) **Calculate** the **turns ratio** of each transformer.

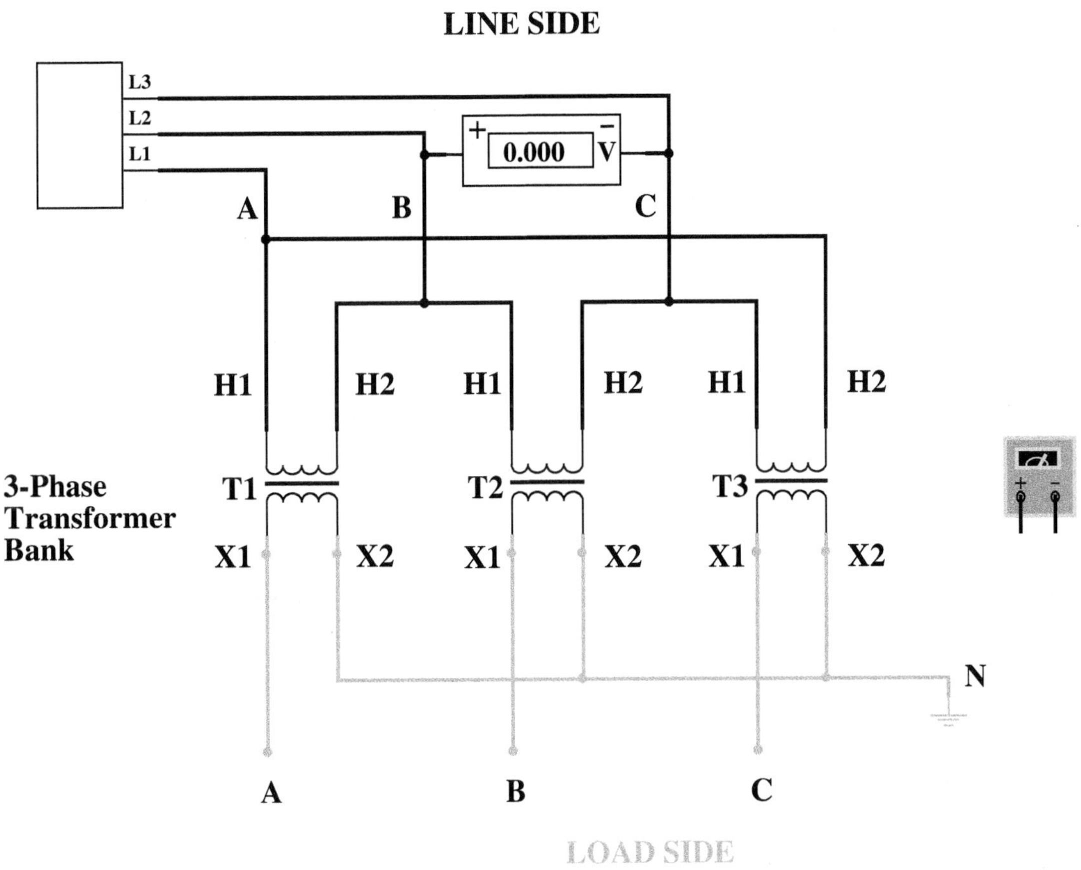

Answers

SECTION 4

Electrical Installation and Maintenance

Chapter 20	Circuit Conductors and Wire Sizes	221
Chapter 21	Fuses and Circuit Breakers	227
Chapter 22	Relays	233
Chapter 23	Lighting Equipment	237
Chapter 24	Electric Motors and Controls	240
Chapter 25	Electronic Controls	241

Chapter 20 Circuit Conductors and Wire Sizes

20-1 Open the Multisim circuit file CH20_01.

(a) **Record** the value of the **source** and **load voltages** with the **load switched OFF.**

(b) **Repeat** part (a) with the **load switched ON.**

(c) **Determine** the value of the **line voltage drop**.

(d) **Calculate** the **percent voltage drop**.

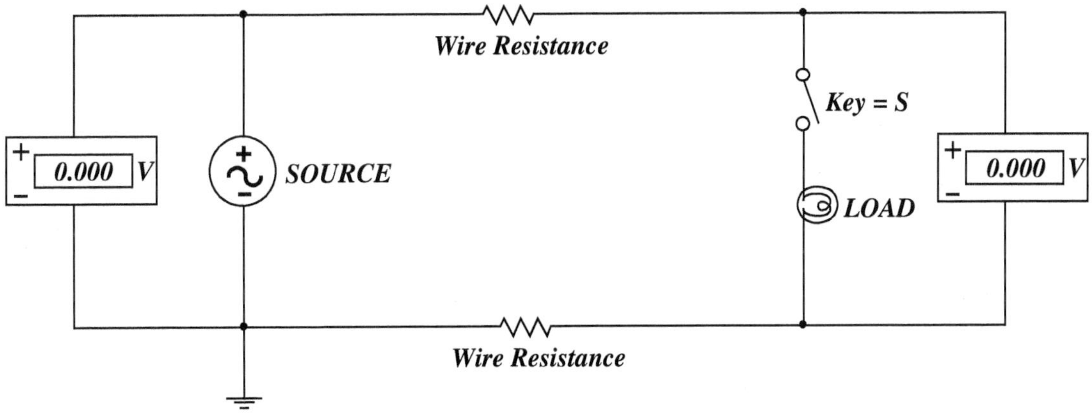

Answers

20-2 Open the Multisim circuit file CH20_02.

(a) **Record** the value of the **load current.**

(b) **Calculate** the **line voltage drop** based on the current and wire resistance.

(c) **Calculate** the value of the **voltage across the load.**

(d) **Measure** and **record** the voltage across the load.

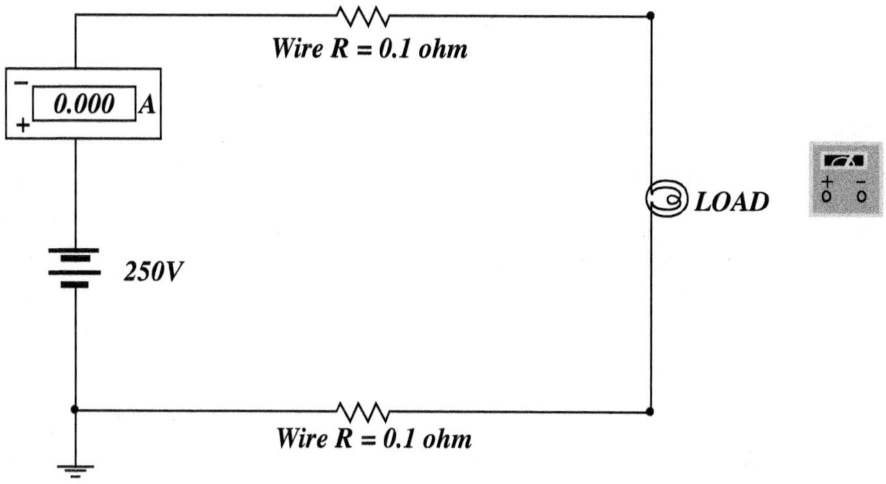

Answers

20-3 Open the Multisim circuit file CH20_03.

(a) **Calculate** the **resistance** of the **1,000 ft of AWG 10** copper using a resistivity **K of 12.9** ohms per circular mil foot.

(b) **Measure** and **record resistance** of the **spool of 1,000 ft. AWG 10 copper wire** shown.

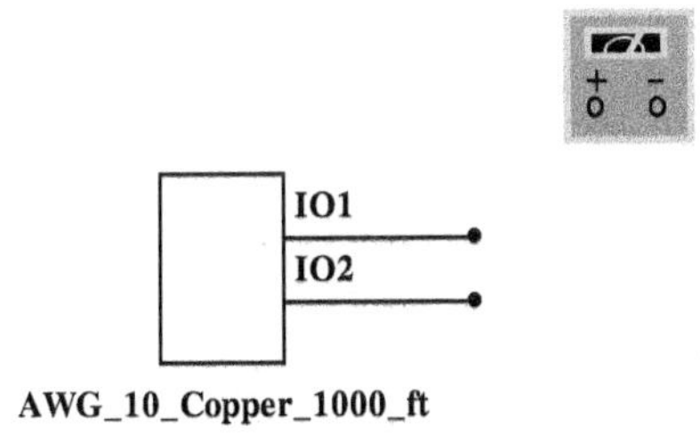

Answers

20-4 Open the Multisim circuit file CH20_04.

(a) **Measure** and **record** the **resistance** between the two wire ends of the roll of **AWG 12 copper wire** shown.

(b) Using a **K of 12.9**, calculate the approximate length of the wire.

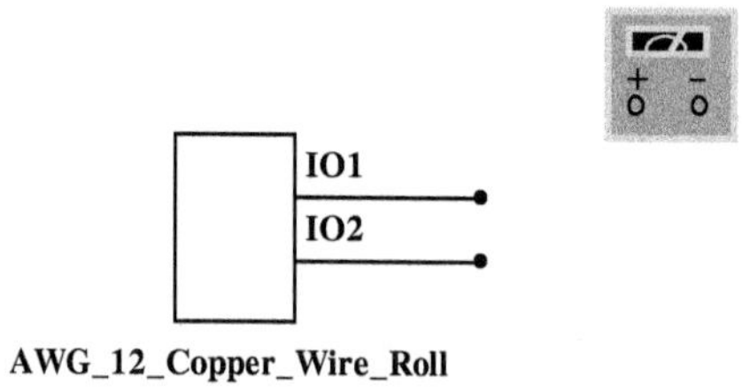

<u>Answers</u>

20-5 Open the Multisim circuit file CH20_05.

(a) **Record** the value of the **load current.**

(b) **Calculate** the **circuit voltage drop** (use a K of 12.9).

(c) **Calculate** the value of the **load voltage**.

(d) **Measure** and **record** the **load voltage.**

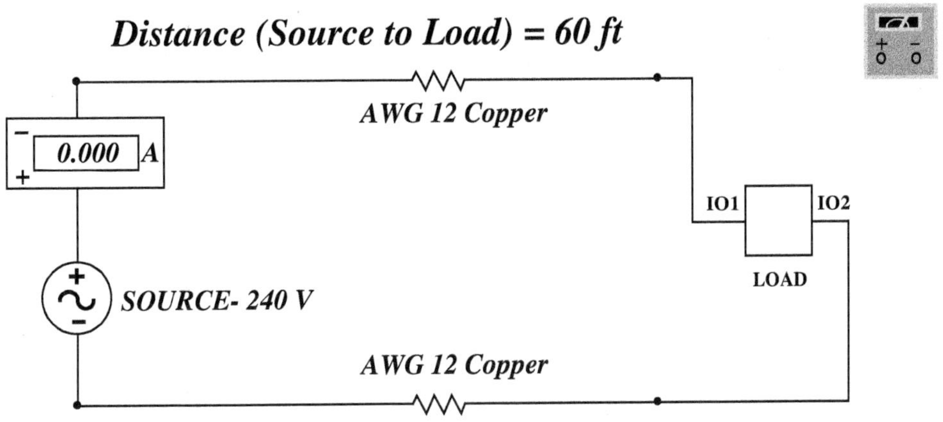

Answers

20-6 Open the Multisim circuit file CH20_06.

(a) **Record** the **voltage of the load.**

(b) **Repeat** with an **additional 50 ft of AWG 12** wire connected in **series** with **each of the line wires** of the original circuit.

(c) **Repeat** with an **additional 50 ft of AWG 12 wire** connected in **parallel** with **each of the line wires** of the original circuit.

(d) Which circuit (**a, b, or c**) results in **the lowest voltage drop**?

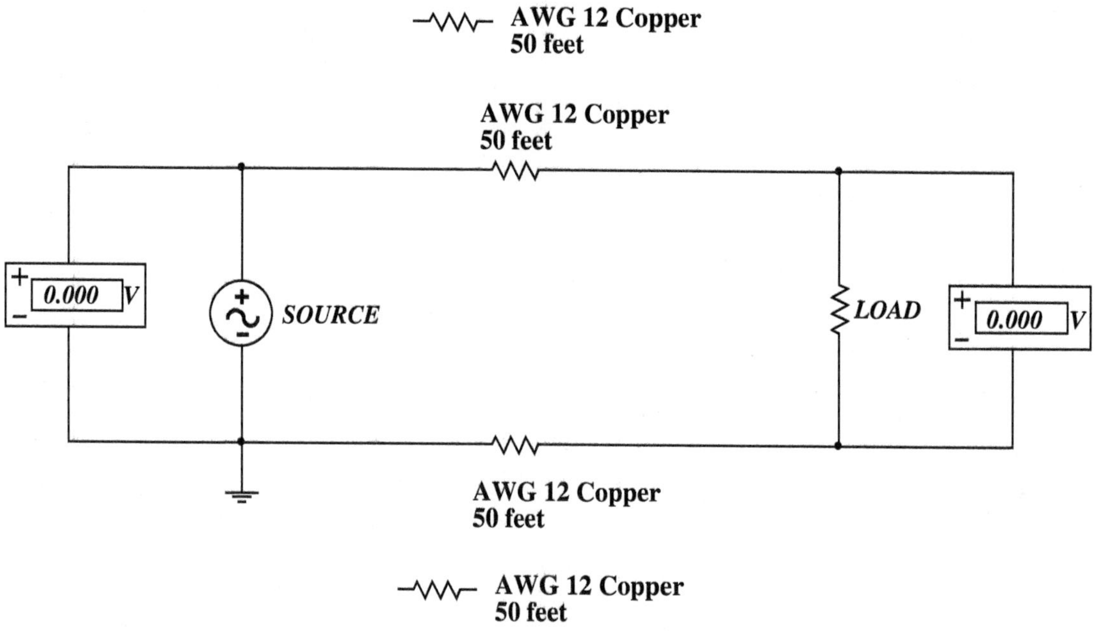

Answers

Chapter 21 Fuses and Circuit Breakers

21-1 Open the Multisim circuit file CH21_01.

(a) **Record** the **current** drawn by **each lamp** load.

(b) Closing which **two switches** will result in an **overload** condition? What is the **value** of this **overload current**?

(c) Which **three switches** can be closed **without creating an overload** condition? What is the **value** of this **current** flow?

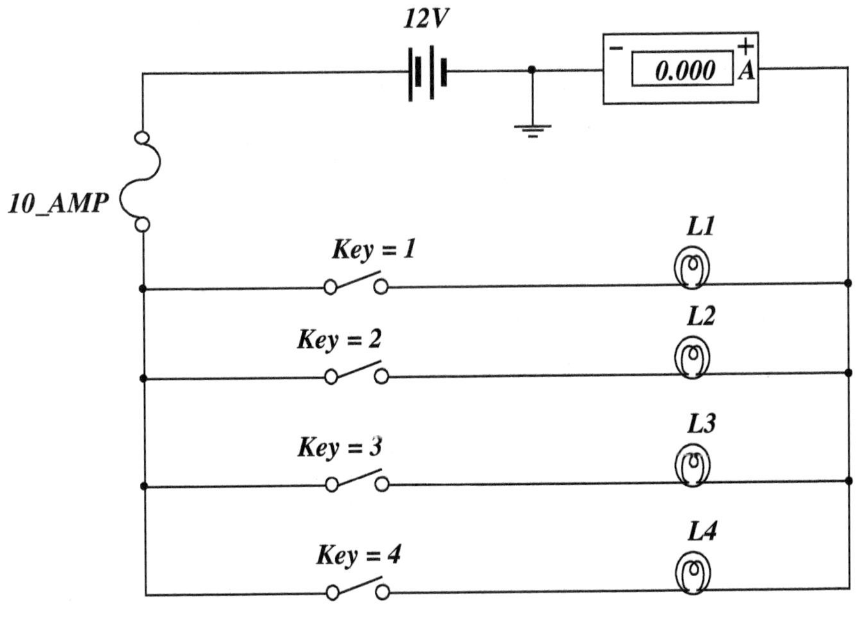

Answers

21-2 Open the Multisim circuit file CH21_02.

Closing **which switch(es)** will create a **short circuit?**

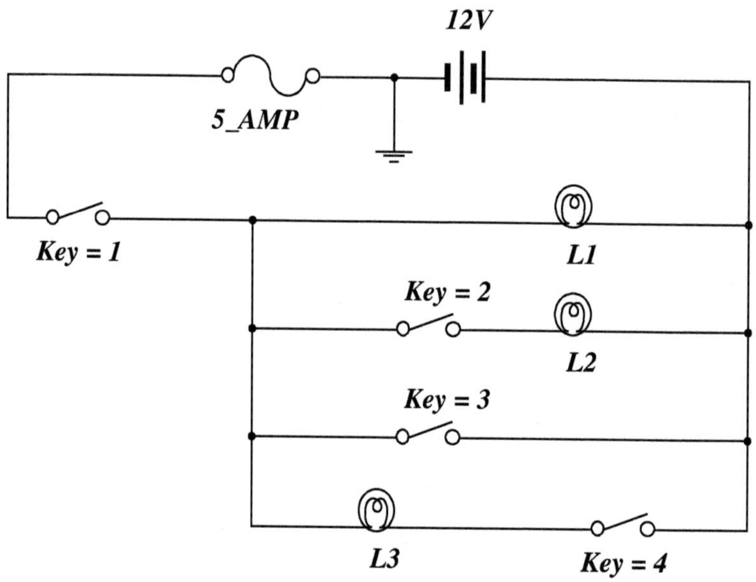

Answer

21-3 Open the Multisim circuit file CH21_03.

(a) **Record** the **voltage** on the **LINE** side of the fuses.

(b) **Record** the **voltage** on the **LOAD** side of the fuses.

(c) **Record** the **voltage** from **L1 to T2.**

(d) **Record** the **voltage** from **L2 to T1**.

(e) Base on your readings, **which fuse is burnt open?** (Replace blown fuse to confirm your answer.)

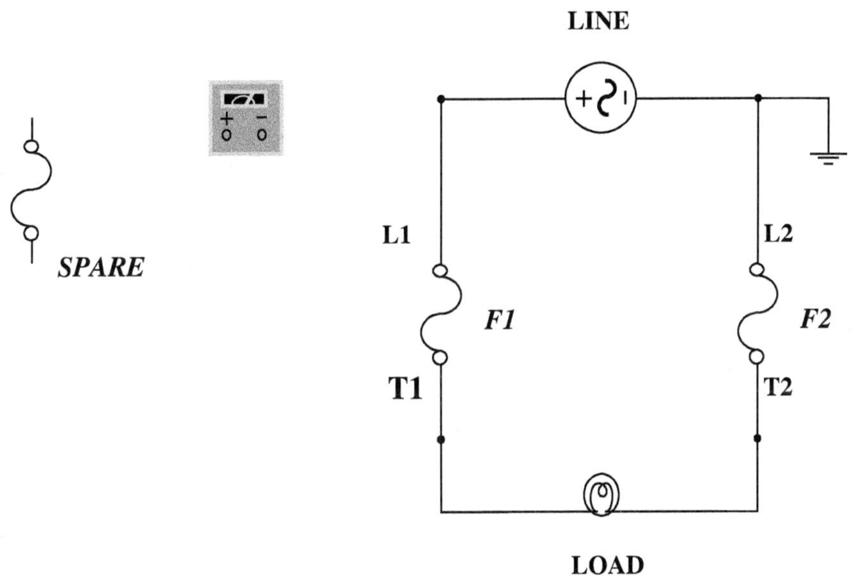

<u>**Answers**</u>

21-4 Open the Multisim circuit file CH21_04.

(a) **Measure** and **record** the resistance of the **GOOD fuse.**

(b) **Measure** and **record** the resistance of the **OPEN fuse.**

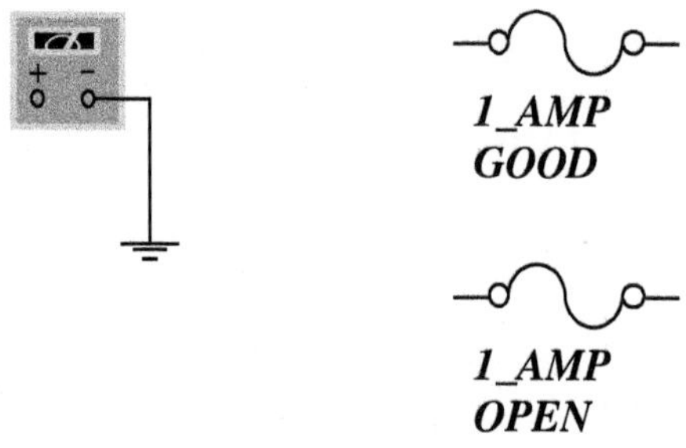

<u>Answer</u>

21-5 Open the Multisim circuit file CH21_05.

Connect the **multimeter** to **determine** which fuses are **good** and which are **burnt open**.

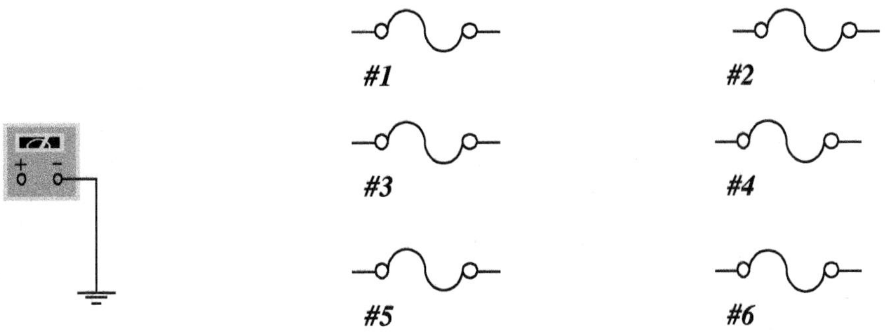

Answers

21-6 Open the Multisim circuit file CH21_06.

(a) **Measure** and **record** the **voltage** drop across the **fuse** for the **normally operating circuit.**

(b) **Measure** and **record** the **voltage** drop across the **fuse** with the **open fuse connected** in place of the good one.

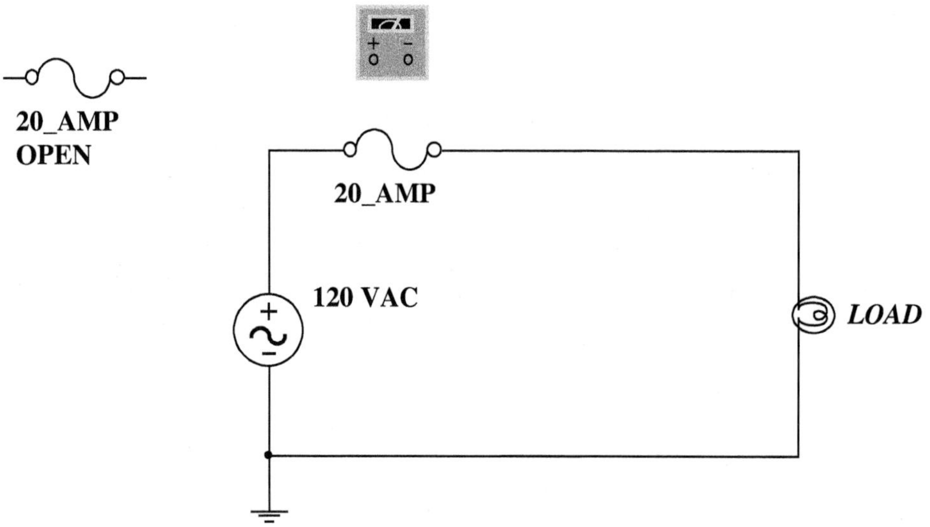

Answers

Chapter 22 Relays

22-1 Open the Multisim circuit file CH22_01.

(a) **Complete the circuit diagram** with the **relay connected** so that the **lamp load circuit** is operated by the **switch control** circuit.

(b) **Connect the Multisim circuit** according to the diagram and **simulate the operation** of the circuit.

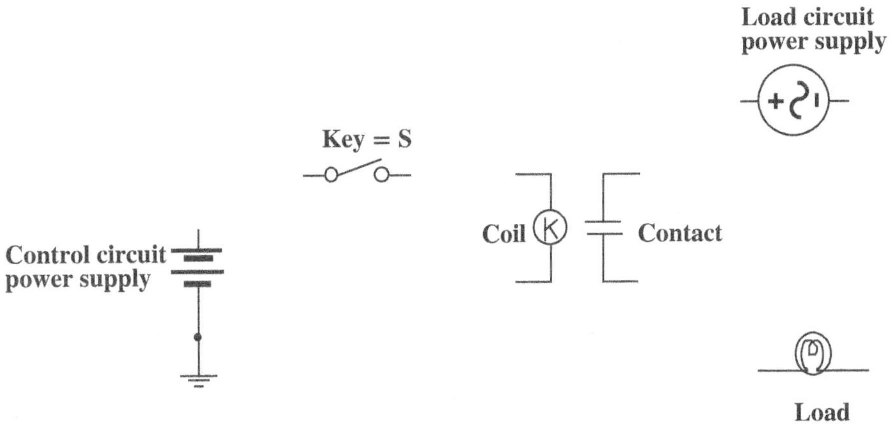

Answer

22-2 Open the Multisim circuit file CH22_02.

Operate the simulated circuit and answer the following:

(a) Which contact is the **normally-open contact**? **Why**?

(b) Which contact is the **normally-closed contact**? **Why**?

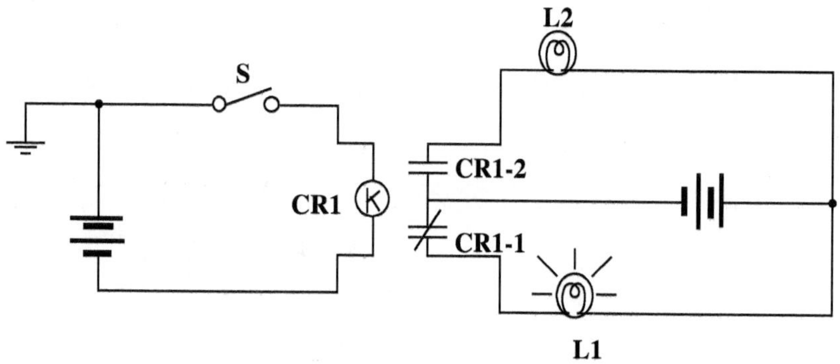

Answers

22-3 Open the Multisim circuit file CH22_03.

(a) **Measure** and **record** the operating **relay coil voltage**.

(b) **Measure** and **record** the operating **load voltage.**

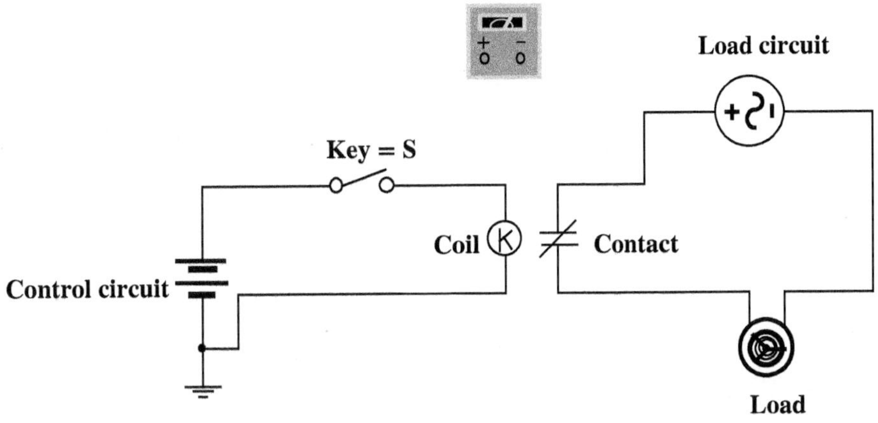

<u>**Answers**</u>

22-4 Open the Multisim circuit file CH22_04.

(a) **Measure** and **record** the relay **coil current**.

(b) **Measure** and **record** the relay **load current.**

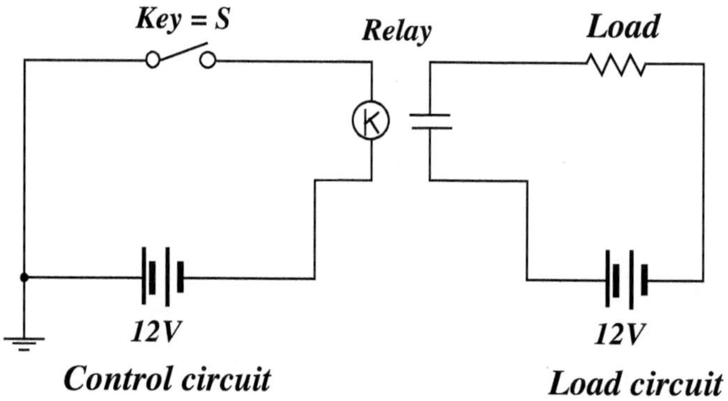

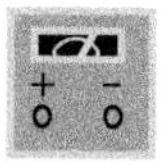

Answers

Chapter 23 Lighting Equipment

23-1 Open the Multisim circuit file CH23_01.

(a) **Complete the circuit diagram** with the **single-pole switch** connected to control the **two lamps from one position.**

(b) **Connect the Multisim circuit** according to the diagram and **simulate the operation** of the circuit.

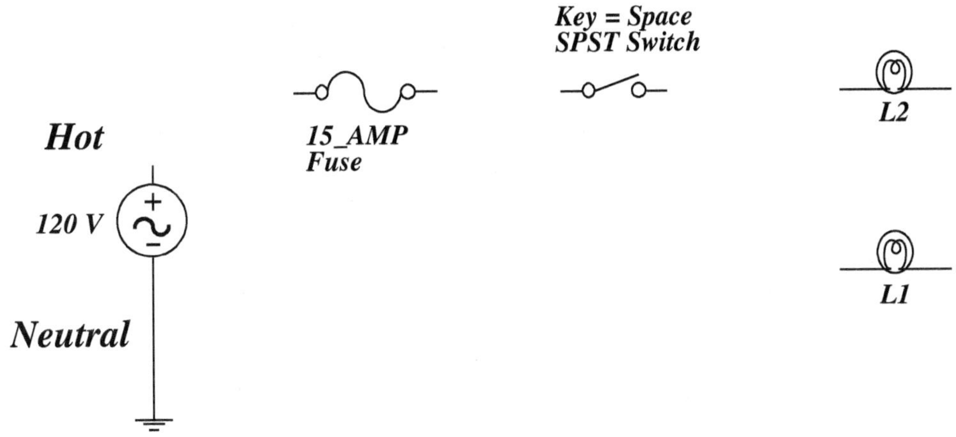

Answer

23-2 Open the Multisim circuit file CH23_02.

(a) **Complete the circuit diagram** with the **two 3-way switches** connected to control the **two lamps from two positions.**

(b) **Connect the Multisim circuit** according to the diagram and **simulate the operation** of the circuit.

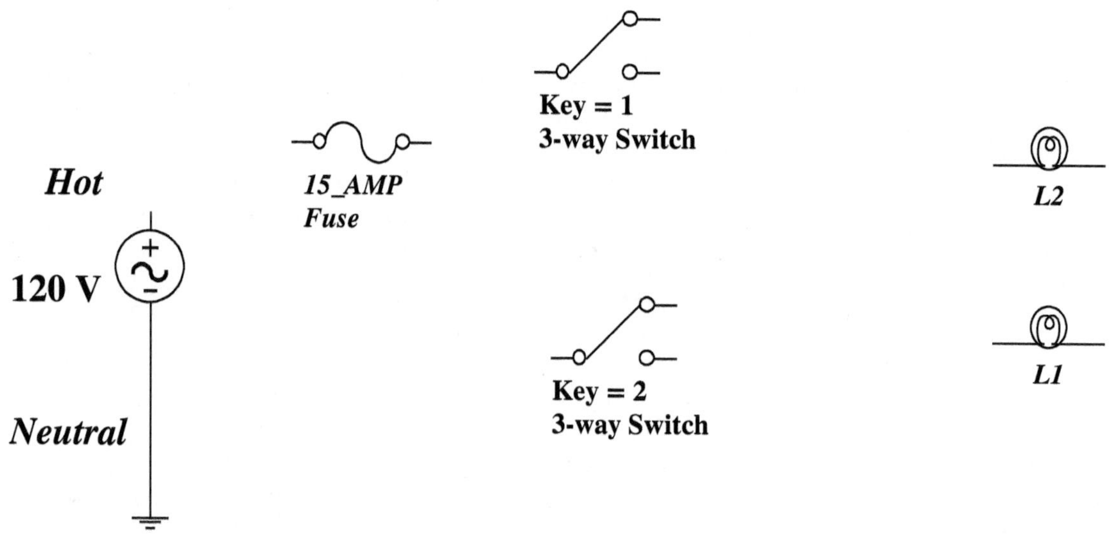

<u>**Answer**</u>

23-3 Open the Multisim circuit file CH23_03.

Operate the lamp controlled from two positions circuit and **identify the wiring problem that exists.**

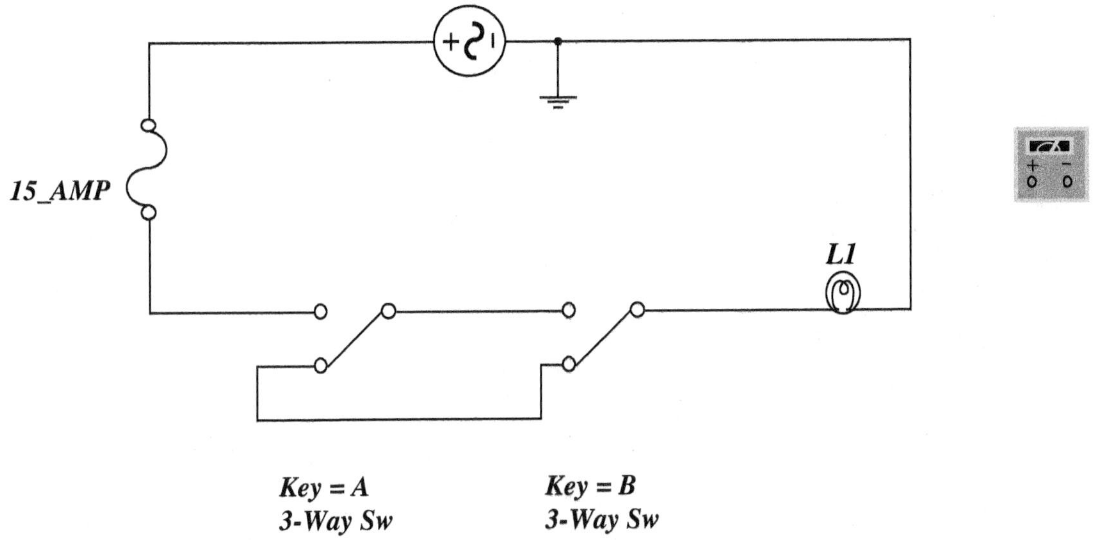

<u>**Answer**</u>

Chapter 24 Electric Motors and Controls

24-1 Open the Multisim circuit file CH24_01.

(a) **Measure** and **record** the value of the **DC control voltage**.

(b) **Measure** and **record** the value of the **3-Phase AC line voltage**.

(c) **Measure** and **record** the value of the **DC starter coil control current.**

(d) **Measure** and **record** the value of the **3-phase AC motor current**.

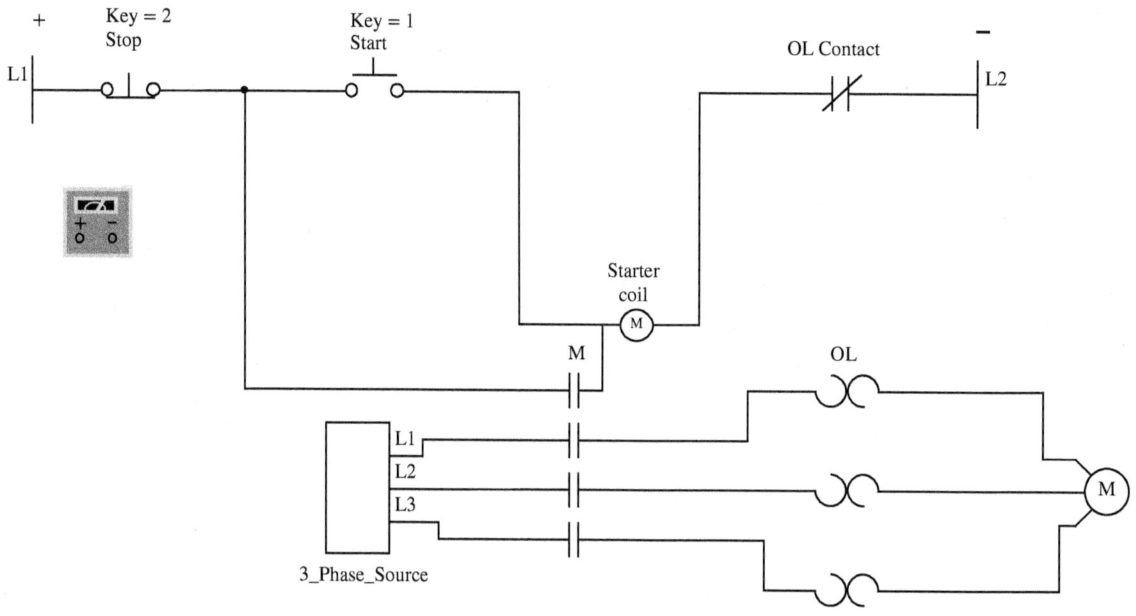

Answers

Chapter 25 Electronic Controls

25-1 Open the Multisim circuit file CH25_01.

(a) **Record** the value of the **current** flow for **circuit #1.** What diode bias connection (**forward or reverse**) is used?

(b) **Record** the value of the **current** flow for **circuit #2.** What diode bias connection (**forward or reverse**) is used?

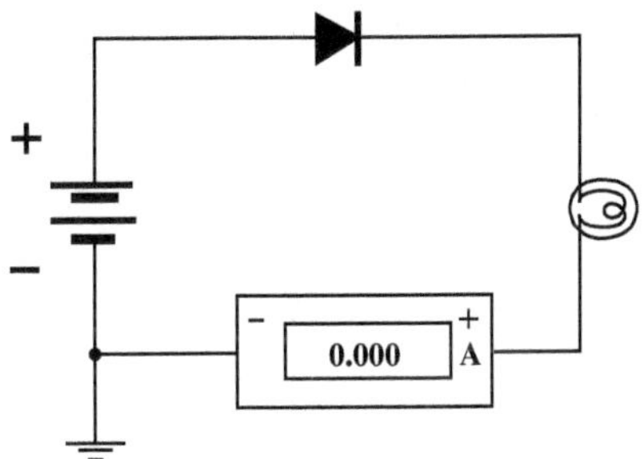

Circuit #1

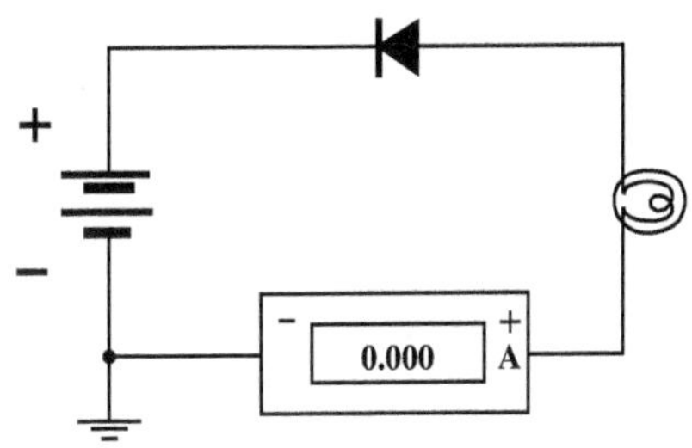

Circuit #2

<u>Answers</u>

25-2 Open the Multisim circuit file CH25_02.

(a) **Complete the circuit diagram** with the **resistor and LED** properly connected in series so as to operate the LED from the DC voltage source.

(b) **Connect the Multisim circuit** according to the diagram and **simulate the operation** of the circuit.

(c) **Measure** and **record** the value of the **LED operating current.**

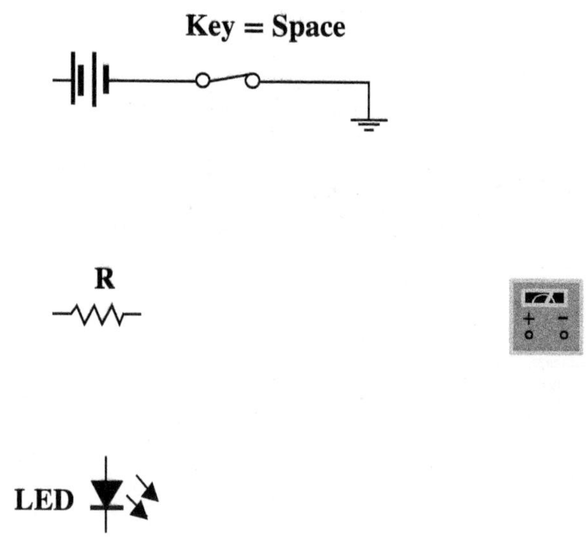

Answers

25-3 Open the Multisim circuit file CH25_04.

(a) **Measure** and **record** the value of the **weak input signal.**

(b) **Measure** and **record** the value of the **amplified output signal.**

(c) Approximately how many times greater in the **output signal as compared to the input signal**?

(d) Examine the input and output wave forms displayed on the scope. What is the phase relationship between the two?

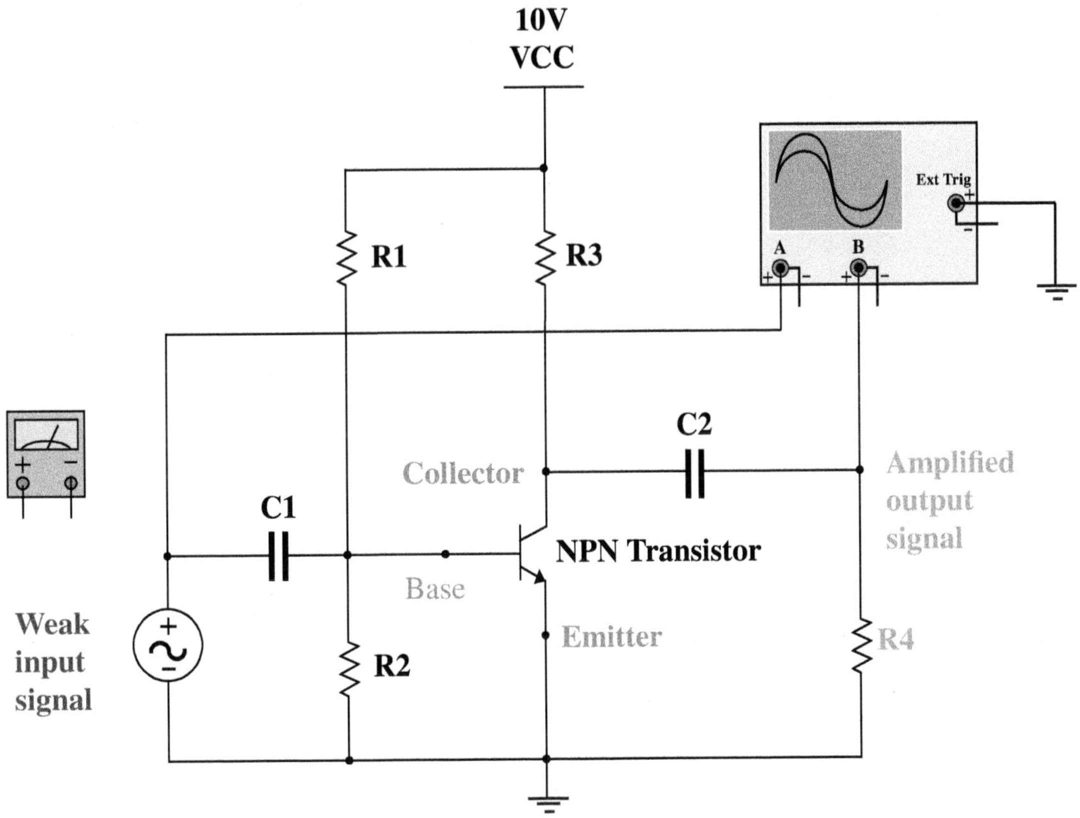

<u>Answers</u>

25-4 Open the Multisim circuit file CH25_04.

(a) What **type of waveform** is seen across the load when the output is **switched ON**?

(b) **Measure** and **record** the **supply voltage**.

(c) **Measure** and **record** the **operating voltage** across the **load**.

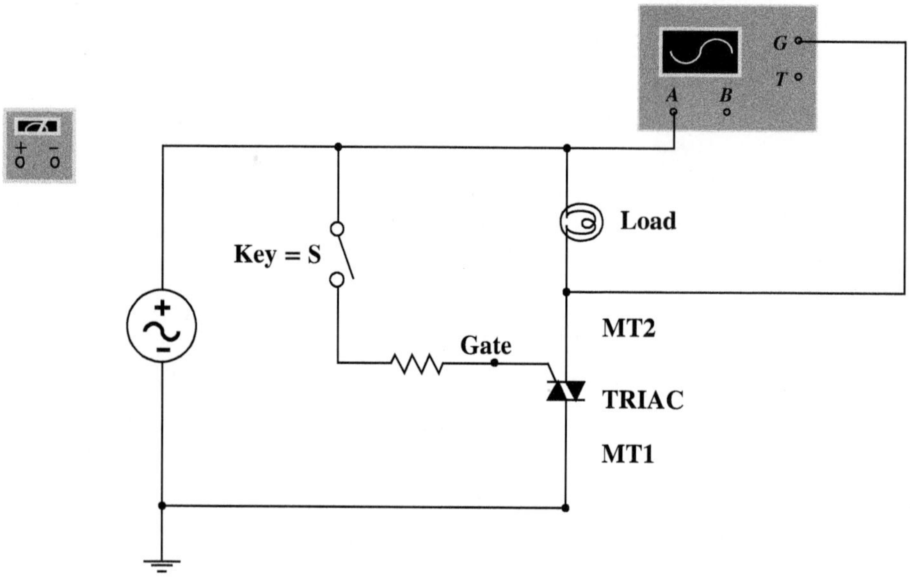

Answers

25-4 Open the Multisim circuit file CH25_04.

(e) **Measure** and **record** the **operating voltage** across the **lamp**.

(f) **Measure** and **record** the **operating current** through the **lamp**.

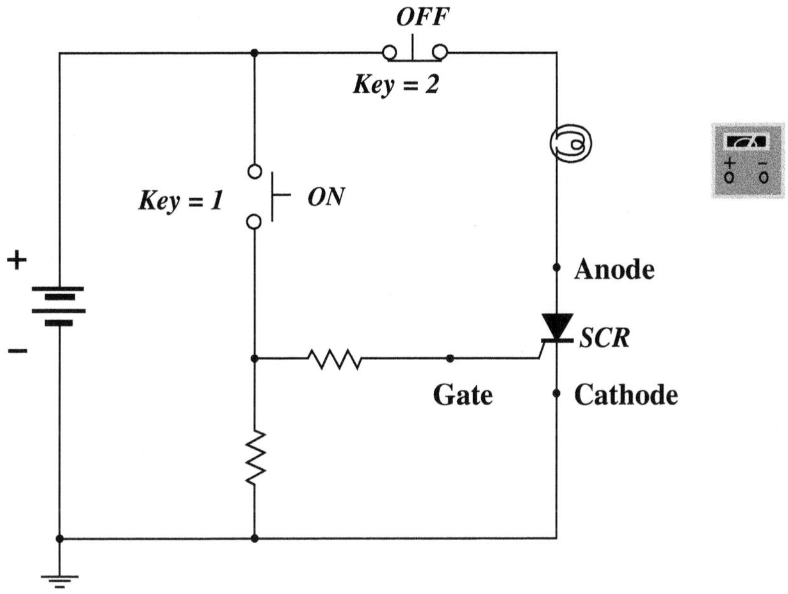

Answers

25-5 Open the Multisim circuit file CH25_05.

(a) What type of rectifier circuit is simulated?

(b) Examine the AC input and DC output waveforms of the oscilloscopes. **In what way are they different?**

(c) **Measure** and **record** the value of the **AC input voltage.**

(d) **Measure** and **record** the value of the **DC output voltage.**

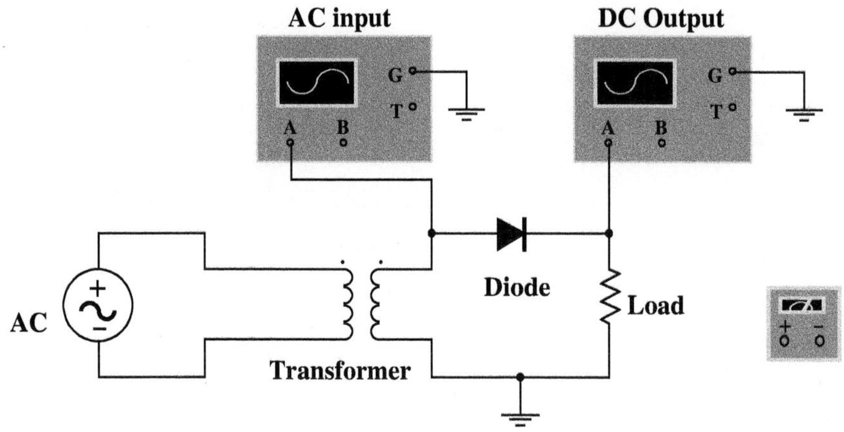

<u>Answers</u>

25-6 Open the Multisim circuit file CH25_06.

(a) What type of rectifier circuit is simulated?

(b) Examine the DC output waveform of the oscilloscope. **In what way is it different from the AC input?**

(c) **Measure** and **record** the value of the **AC input voltage.**

(d) **Measure** and **record** the value of the **DC output voltage.**

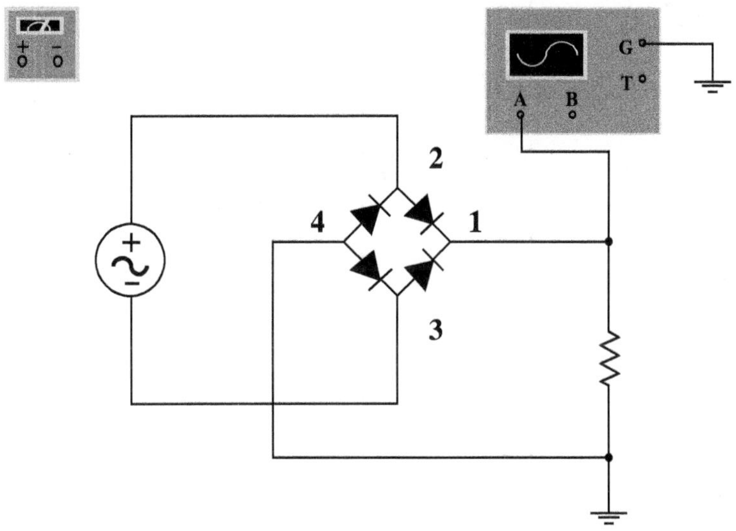

Answers

25-7 Open the Multisim circuit file CH25_07.

Operate the **transistor switching** circuit.

Measure and **record**:

(a) the **sensor circuit** DC **voltage** and **current**

(b) the **relay control circuit** DC **voltage** and **current**

(c) the AC **load voltage** and **current**

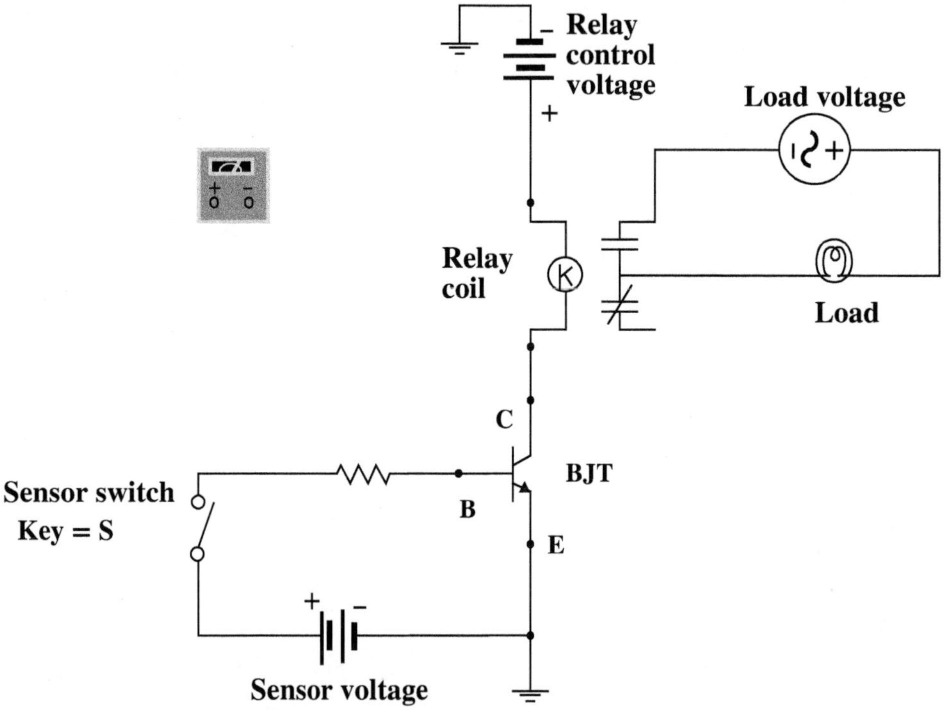

Answers

25-8 Open the Multisim circuit file CH25_08.

(a) Measure and record the value of the **weak input signal.**

(b) Measure and record the value of the **amplified output signal**.

(c) Approximately how many times greater in the **output signal as compared to the input signal**?

(d) Examine the input and output wave forms displayed on the scope. What is the **phase** relationship between the two?

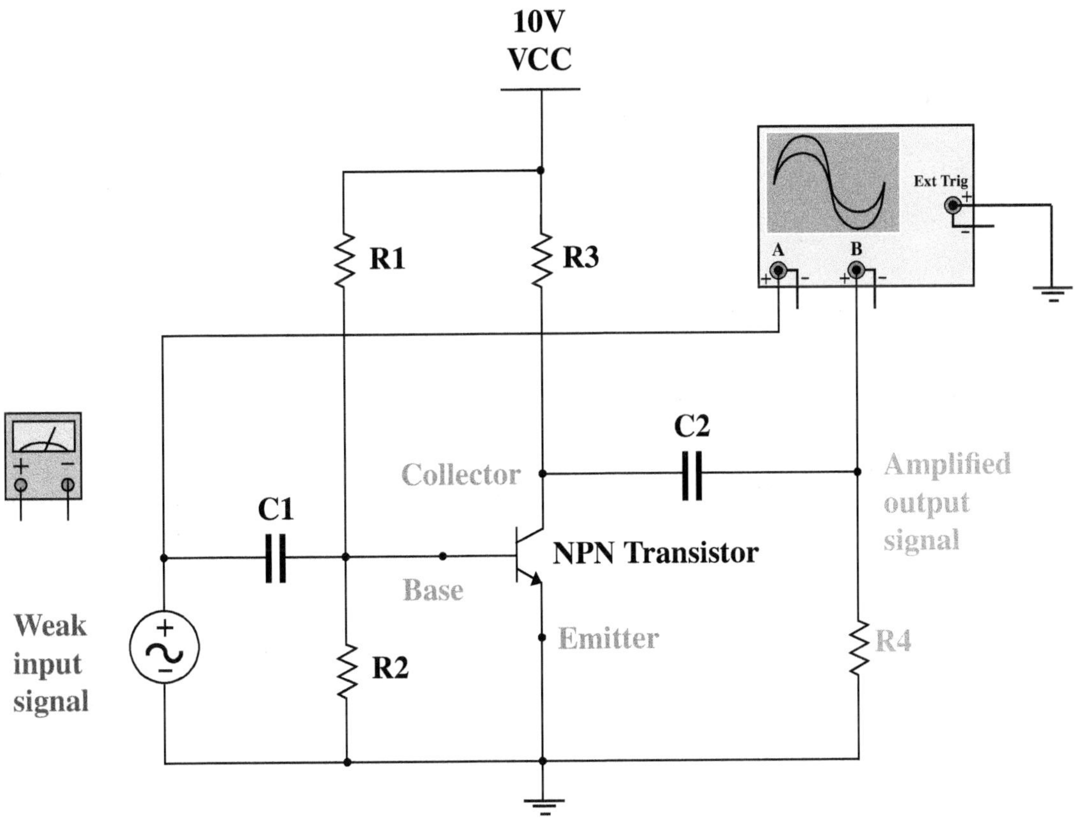

Answers

25-9 Open the Multisim circuit file CH25-09

Generate a digital signal waveform by toggling the switch between the HIGH and LOW positions. Record the High and Low values of the digital signal generated (scope setting 5 Volts per Division).

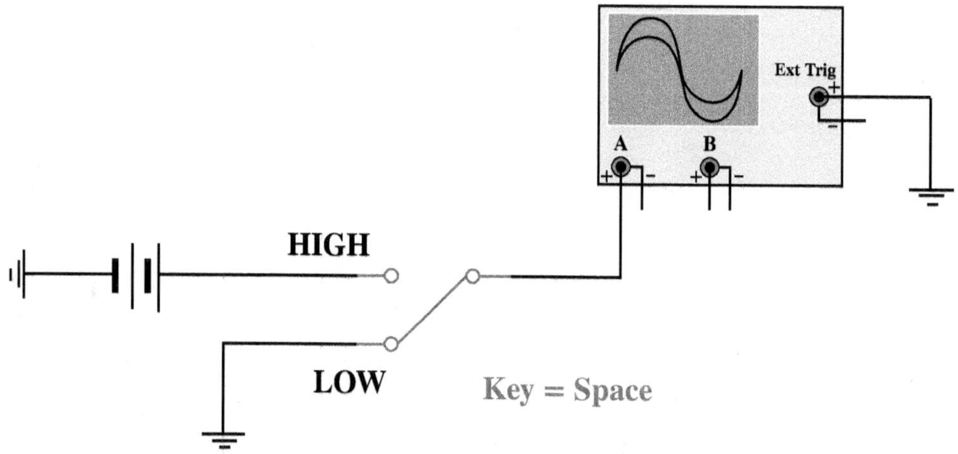

Answers

25-10 Open the Multisim circuit file CH25-10

Operate the 2-input AND gate circuit shown.

Construct a truth table for the circuit by observing the status of the logic probes.

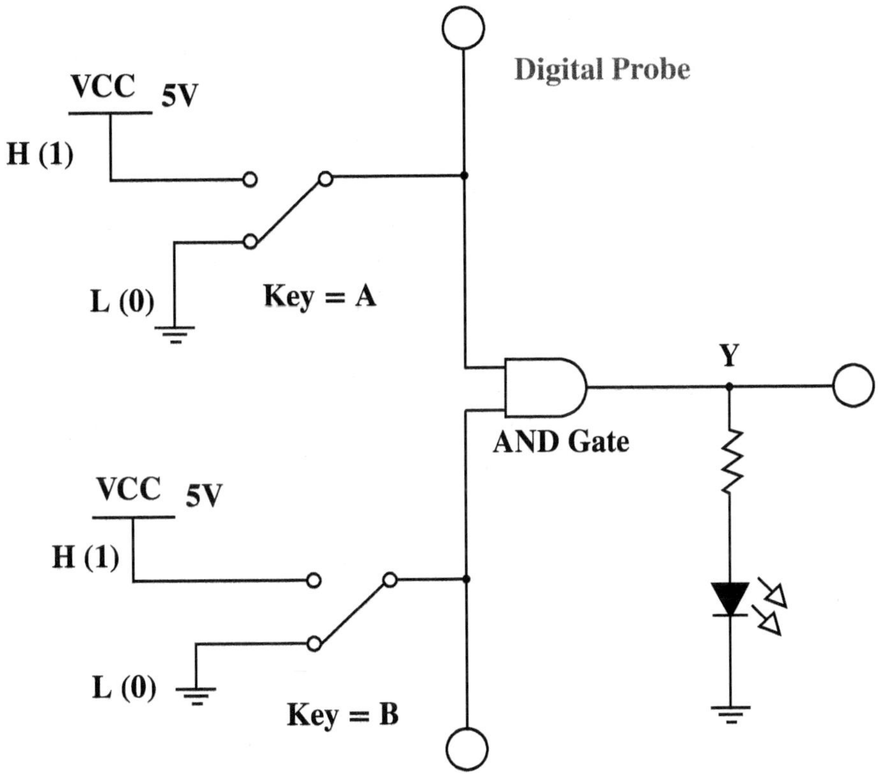

Answers

25-11 Open the Multisim circuit file CH25-11

Operate the 2-input OR gate circuit shown.

Construct a truth table for the circuit by observing the status of the logic probes.

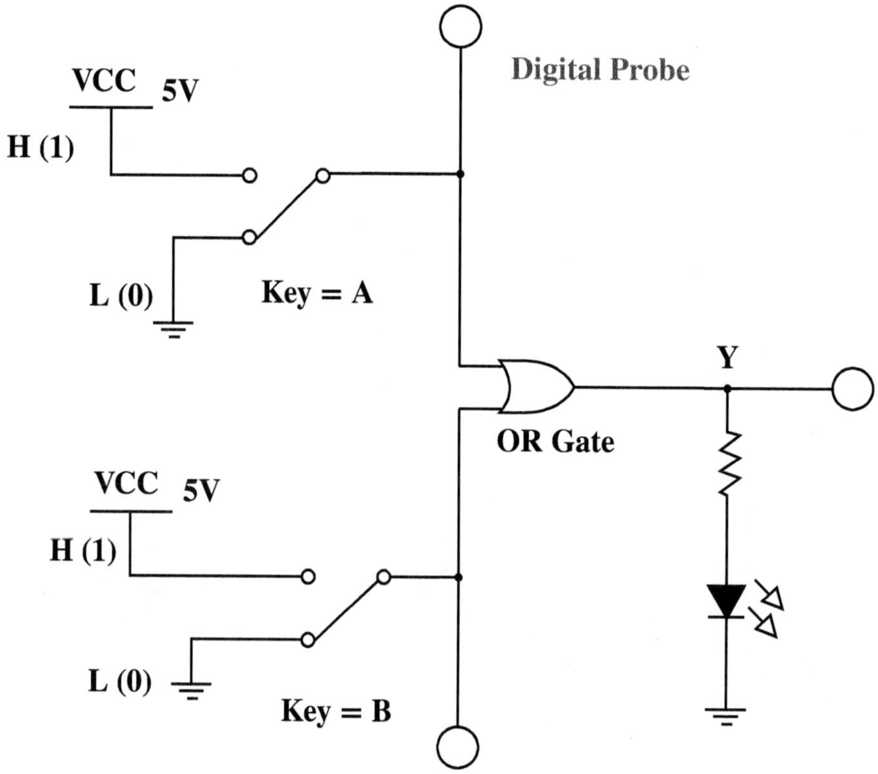

Answers

25-12 Open the Multisim circuit file CH25-12

Operate the NOT (Inverter) gate circuit shown.

Construct a truth table for the circuit by observing the status of the logic probes.

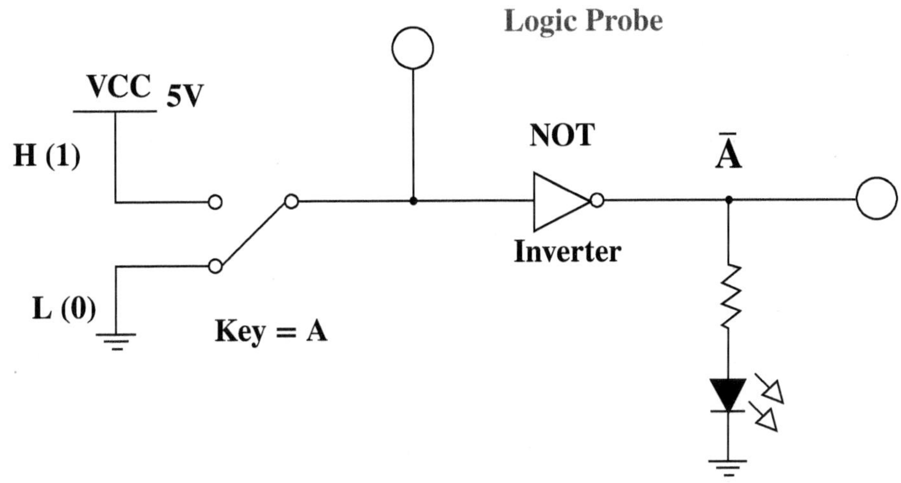

Answers

25-13 Open the Multisim circuit file CH25-13

Operate the 2-input NAND gate circuit shown.

Construct a truth table for the circuit by observing the status of the logic probes.

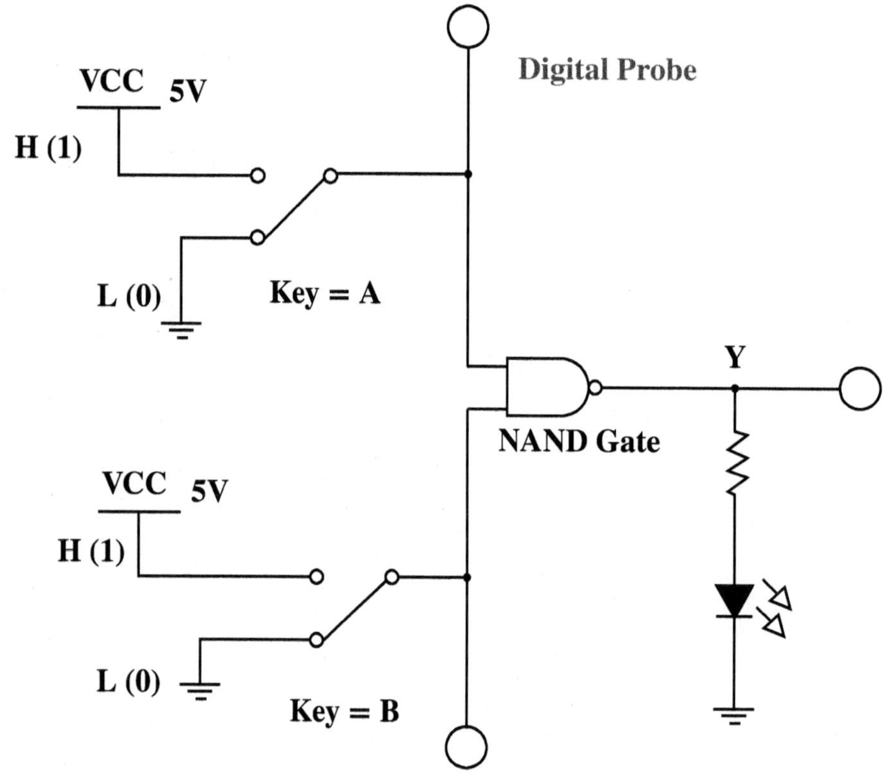

Answers

25-14 Open the Multisim circuit file CH25-14

Operate the 2-input NOR gate circuit shown.
Construct a truth table for the circuit by observing the status of the logic probes.

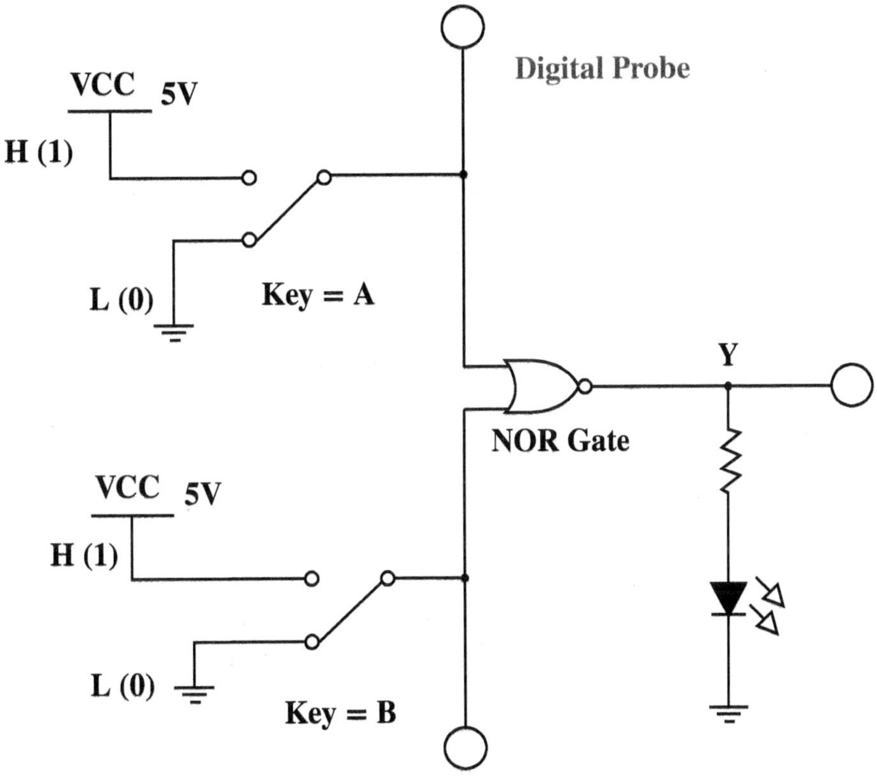

Answers

25-15 Open the Multisim circuit file CH25-15

Operate the XOR gate circuit shown.

Construct a truth table for the circuit by observing the status of the logic probes.

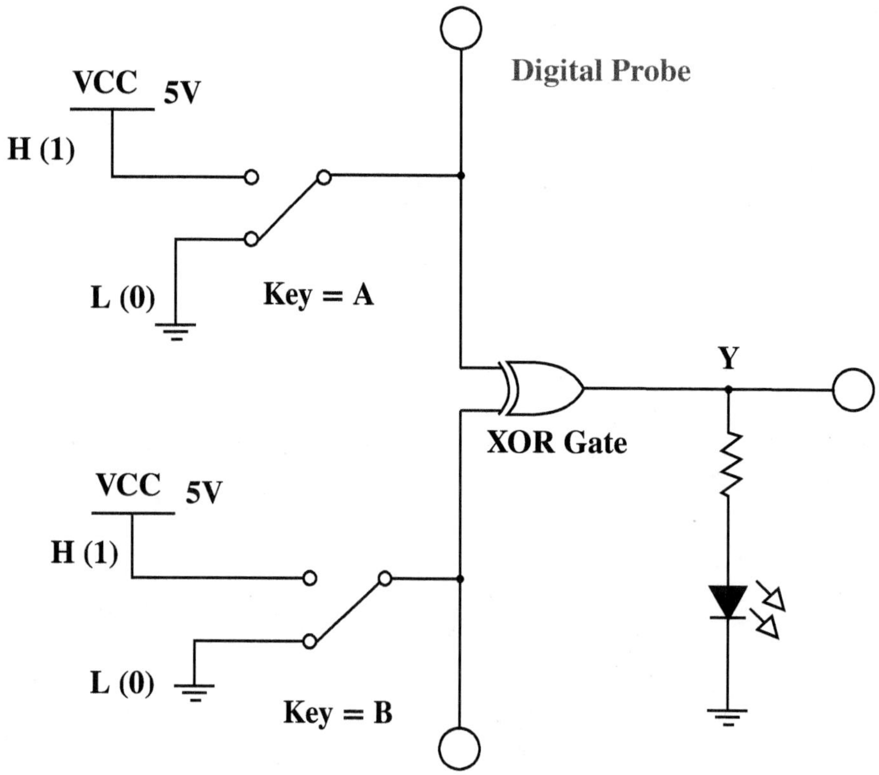

Answers

25-16 Open the Multisim circuit file CH25-16

Operate the combination gate circuit shown.

Construct a truth table for the circuit by observing the status of the logic probes.

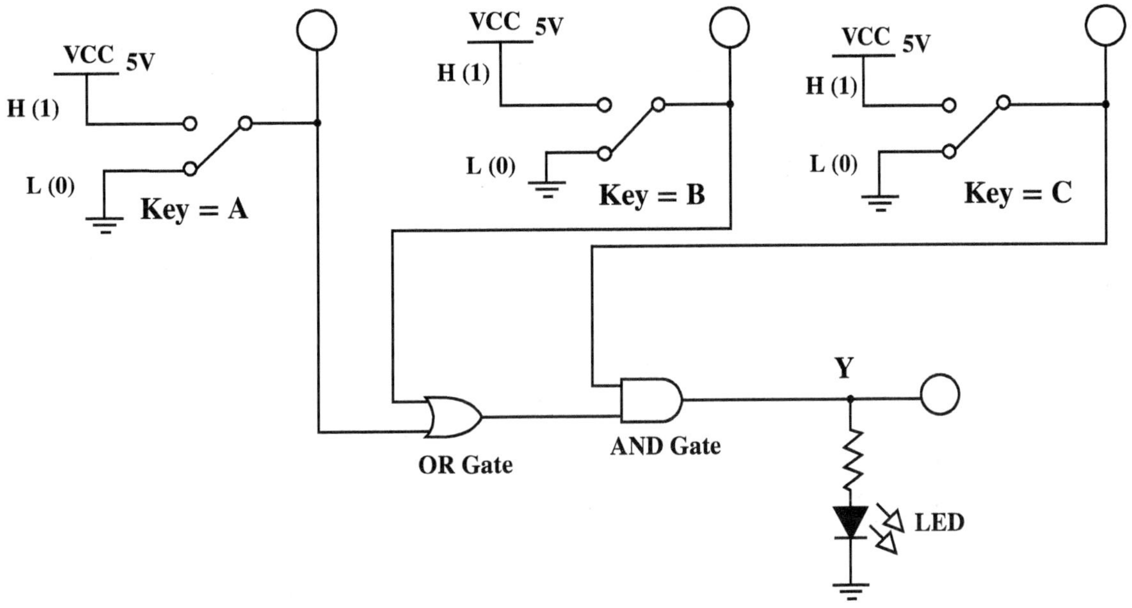

Answers

25-17 Open the Multisim circuit file CH25-17

Operate the combination gate circuit shown.

Construct a truth table for the circuit by observing the status of the logic probes.

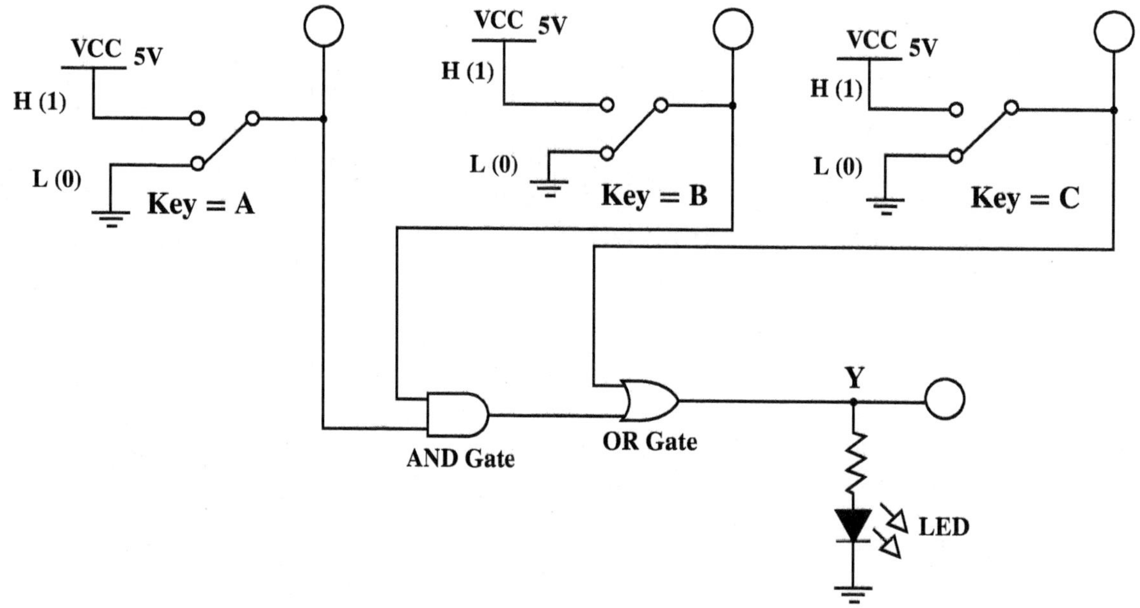

Answers

NOTES

NOTES